Undergraduate Texts in Mathematics

Editors
S. Axler
F.W. Gehring
P.R. Halmos

Springer
New York
Berlin
Heidelberg
Barcelona
Budapest
Hong Kong
London
Milan
Paris
Santa Clara
Singapore
Tokyo

Undergraduate Texts in Mathematics

Judith N. Cederberg

A Course in
Modern Geometries

Springer

Judith N. Cederberg
Department of Mathematics,
St. Olaf College,
Northfield, MN 55057
USA

S. Axler
Department of Mathematics,
Michigan State University,
East Lansing, MI 48824
USA

F.W. Gehring
Department of Mathematics,
University of Michigan,
Ann Arbor, MI 48109
USA

P.R. Halmos
Department of Mathematics,
Santa Clara University,
Santa Clara, CA 95053
USA

Mathematics Subject Classification (1991): 51-XX

Library of Congress Cataloging-in-Publication Data

Cederberg, Judith N.
    A course in modern geometries / by Judith N. Cederberg.
        p.   cm.
    Bibliography: p.
    Includes index.
    1. Geometry. I. Title.
QA445.C36   1989
516—dc19

88-35478
CIP

Typeset by Thomson Press,   New Delhi, India
Printed and bound by R.R. Donnelley & Sons, Harrisonburg, Virginia
Printed in the United States of America.

9 8 7 6 5 4 3 (Corrected third printing, 1995)

ISBN 0-387-96922-5 Springer-Verlag New York Berlin Heidelberg
ISBN 3-540-96922-5 Springer-Verlag Berlin Heidelberg New York

To Jim,
Anna, and Rachel

Preface

The origins of geometry are lost in the mists of ancient history, but geometry was already the preeminent area of Greek mathematics over 20 centuries ago. As such, it became the primary subject of Euclid's *Elements*. *Elements* was the first major example of a formal axiomatic system and became a model for mathematical reasoning. However, the eventual discoveries of non-Euclidean geometries profoundly affected both mathematical and philosophical understanding of the nature of mathematics. The relation between Euclidean and non-Euclidean geometries became apparent with the development of projective geometry—a geometry with origins in artists' questions about perspective.

This interesting historical background and the major philosophical questions raised by developments in geometry are virtually unknown to current students, who often view geometry as a dead subject full of two-column proofs of patently clear results. It is no surprise that Mary Kantowski, in an article entitled "Impact of Computing on Geometry," has called geometry "the most troubled and controversial topic in school mathematics today" (Fey, 1984, p. 31). However, this and many other recent articles provide evidence for an increasing realization that the concepts and methods of geometry are becoming more important than ever in this age of computer graphics. The geometry of the artists, projective geometry, has become the tool of computer scientists and engineers as they work on the frontiers of CAD/CAM (computer-aided design/computer-aided manufacturing) technology.

The major emphasis of this text is on the geometries developed after Euclid's *Elements* (circa 300 B.C.). In addition to the primary goal of studying these "newer" geometries, this study provides an excellent opportunity to explore aspects of the history of mathematics. Also, since algebraic techniques are frequently used, this study demonstrates the interaction of several areas of mathematics and serves to develop geometrical insights into mathematical results that previously appeared to be completely abstract in nature.

Since Euclid's geometry is historically the first major example of an axiomatic system and since one of the major goals of teaching geometry in high school is to expose students to deductive reasoning, Chapter 1 begins with a general description of axiomatic (or deductive) systems. Following this general introduction, several finite geometries are presented as examples of specific systems. These finite geometries not only demonstrate some of the concepts that occur in the geometries of Chapters 2 through 4 but also indicate the breadth of geometrical study.

In Chapter 2, Euclid's geometry is first covered in order to provide historical and mathematical preparation for the major topic of non-Euclidean geometries. This brief exposure to Euclid's system serves both to recall familiar results of Euclidean geometry and to show how few substantial changes have occurred in Euclidean geometry since Euclid formulated it. The non-Euclidean geometries are then introduced to demonstrate that these geometries, which appear similar to Euclidean geometry, have properties that are radically different from comparable Euclidean properties.

The beginning of Chapter 3 serves as a transition from the synthetic approach of the previous chapters to the analytic treatment contained in the remainder of this chapter and the next. There follows a presentation of Klein's definition of geometry, which emphasizes geometrical transformations. The subsequent study of the transformations of the Euclidean plane begins with isometries and similarities and progresses to the more general transformations called affinities.

By using an axiomatic approach and generalizing the transformations of the Euclidean plane, Chapter 4 offers an introduction to projective geometry and demonstrates that this geometry provides a general framework within which the geometries of Chapters 2 and 3 can be placed.

Although the text ends here, mathematically the next logical step in this process is the study of topology, which is usually covered in a separate course.

This text is designed for college-level survey courses in geometry. Many of the students in these courses are planning to pursue secondary-school teaching. However, with the renewed interest in geometry, other students interested in further work in mathematics or computer science will find the background provided by these courses increasingly valuable. These survey courses can also serve as an excellent vehicle for demonstrating the relationships between mathematics and other liberal arts disciplines. In an attempt to encourage student reading that further explores these relationships, each chapter begins with a section that lists suggested bibliographic sources for relevant topics in art, history, applications, and so on. I have found that having groups of students research and report on these topics not only introduces them to the wealth of expository writing in mathematics but also provides a way to share their acquired insights into the liberal arts nature of mathematics.

The material contained in this text is most appropriate for junior or senior mathematics majors. The only geometric prerequisite is some familiarity with the most elementary high-school geometry. Since the text makes frequent use of matrix algebra and occasional references to more general concepts of linear algebra, a background in elementary linear algebra is helpful. Because the text introduces the concept of a group and explores properties of geometric transformations, a course based on this text provides excellent preparation for the standard undergraduate course in abstract algebra.

I am especially grateful for the patient support of my husband and the general encouragement of my colleagues in the St. Olaf Mathematics Department. In particular, I wish to thank our department chair, Theodore Vessey, for his support and our secretary, Donna Brakke, for her assistance. I am indebted to the many St. Olaf alumni of Math 80 who studied from early drafts of the text and to Charles M. Lindsay for his encouragement after using preliminary versions of the text in his courses at Coe College in Cedar Rapids, Iowa. Others who used a preliminary version of the text and made helpful suggestions are Thomas Q. Sibley of St. John's University in Collegeville, Minnesota, and Martha L. Wallace of St. Olaf College. I am also indebted to Joseph Malkevitch of York College of the City University of New York for serving as mathematical reader for the text, and to Christina Mikulak for her careful editorial work.

Changes in this printing:
This printing contains a few corrections and changes made for clarification. Many of these have been made as a result of the kind suggestions of others. In particular, I am most grateful to David Flesner (Gettysburg College), Ockle Johnson (Keene State College), and Myrtle Lewin (Agnes Scott College) for their careful reporting of corrections and suggestions for changes.

Supplements to text:
I have developed a series of computer labs to be used in conjunction with this text.[1] One of these labs makes use of *NonEuclid*.[2] The other labs use *The Geometer's Sketchpad* (© Key Curriculum Press). Directions for these labs are formatted in LaTeX. Both the directions and BinHex files of the pre-made sketches for the Macintosh version of *Sketchpad* are available from me via internet at the address below. Eventually, I hope to have versions of the labs using *Cabri Geometry II*[3] (distributed in the U.S. by Texas Instruments Incorporated).

Judith N. Cederberg
cederj @ stolaf. edu

[1] Developed under the auspices of the St. Olaf FIPSE funded project, "Materials Development for Advanced Computing in the Undergraduate Curriculum."

[2] A freeware program developed at CRPCE of Rice University, Written by Joe Austin, Joel Castellanos, Ervan Darnell, Maria Estrado. © Rice University.

[3] Cabri Geometry II is a trademark of Université Joseph Fourier.

Contents

CHAPTER 1

Axiomatic Systems and Finite Geometries

1.1. Gaining Perspective

Finite geometries were developed in the late 19th century, in part to demonstrate and test the axiomatic properties of "completeness," "consistency," and "independence." They are introduced in this chapter to fulfill this historical role and to develop both an appreciation for and an understanding of the revolution in mathematical and philosophical thought brought about by the development of non-Euclidean geometry. In addition, finite geometries provide relatively simple axiomatic systems in which we can begin to develop the skills and techniques of geometric reasoning. The finite geometries introduced in Sections 1.3 and 1.5 also illustrate some of the fundamental properties of non-Euclidean and projective geometry.

Even though finite geometries were developed as abstract systems, mathematicians have applied these abstract ideas in designing statistical experiments using Latin squares and in developing error-correcting codes in computer science. Section 1.4 develops a simple error-correcting code and shows its connection with finite projective geometries. The application of finite affine geometries to the building of Latin squares is equally intriguing. Since Latin squares are clearly described in several readily accessible sources, the reader is encouraged to explore this topic by consulting the resources listed at the end of this chapter.

1.2. Axiomatic Systems

The study of any mathematics requires an understanding of the nature of deductive reasoning, and geometry has been singled out for introducing this methodology to secondary-school students. There are important historical reasons for choosing geometry to fulfill this role, but these reasons are seldom revealed to secondary-school initiates. This section introduces the terminology essential for a discussion of deductive reasoning so that the extraordi-

1

nary influence of the history of geometry on the modern understanding of deductive systems will become evident.

Deductive reasoning takes place in the context of an organized logical structure called an *axiomatic* (or *deductive*) system. Such a system consists of the following components:

1. Undefined terms.
2. Defined terms.
3. Axioms.
4. A system of logic.
5. Theorems.

Undefined terms are included since it is not possible to define all terms without resorting to circular definitions. In geometrical systems these undefined terms frequently, but not necessarily, include "point," "line," "plane," and "on." Defined terms are not actually necessary, but in nearly every axiomatic system certain phrases involving undefined terms are used repeatedly. Thus it is more efficient to substitute a new term, that is, a defined term, for each of these phrases whenever they occur. For example, in Euclidean geometry we substitute the term "parallel lines" for the phrase "lines which do not intersect." Furthermore, it is impossible to prove all statements constructed from the defined and undefined terms of the system without circular reasoning, just as it is impossible to define all terms. So an initial set of statements is accepted without proof. The statements that are accepted without proof are known as *axioms*. From the axioms, other statements can be deduced or proved using the rules of inference of a system of logic (usually Aristotelian). These latter statements are called *theorems*.

As noted earlier, the axioms of a system must be statements constructed using the terms of the system. But they cannot be arbitrarily constructed since an axiom system must be consistent.

Definition 1.1. An axiomatic system is said to be *consistent* if there do not exist in the system any two axioms, any axiom and theorem, or any two theorems that contradict each other.

It should be clear that it is essential that an axiomatic system be consistent since a system in which both a statement and its negation can be proved is worthless. However, it soon becomes evident that it would be difficult to verify consistency directly from this definition since all possible theorems would have to be considered. Instead, models are used for establishing consistency. A *model* of an axiomatic system is obtained by assigning interpretations to the undefined terms so as to convert the axioms into true statements in the interpretations. If the model is obtained by using interpretations that are objects and relations adapted from the real world, we say we have established *absolute consistency*. In this case, statements corresponding to any contradictory theorems would lead to contradictory statements in the model, but contradictions in the real world are supposedly impossible. On the other hand, if the interpretations assigned are taken from another axiomatic system, we have only tested consistency relative to the consistency of the second axiomatic system; that is, the system we are testing is consistent only if the system within which the interpretations are assigned is consistent. In this second case, we say we have demonstrated *relative consistency* of the first axiomatic system. Because of the number of elements in many axiomatic systems, relative consistency is the best we are able to obtain. We illustrate the use of models to determine consistency of the axiomatic system for four-point geometry.

Axioms for Four-Point Geometry

Undefined Terms. Point, line, on.
Axiom 1. There exist exactly four points.
Axiom 2. Two distinct points are on exactly one line.
Axiom 3. Each line is on exactly two points.

Before demonstrating the consistency of this system, it may be helpful to make some observations about these three statements, which will also apply to other axioms in this text. Axiom 1 explicitly guarantees the existence of exactly four points. However, even though lines are mentioned in Axioms 2 and 3, we cannot ascertain whether or not lines exist until theorems verifying this are proved since there is no axiom that explicitly insures their existence. This is true even though in this system the proof of the existence of lines is almost immediate. Axioms 2 and 3 like many mathematical statements are disguised "if...then" statements. Axiom 2 should be interpreted as follows: If two distinct points exist, then these two points are on exactly one line. Similarly, Axiom 3 should be interpreted: If there is a line, it is on exactly two points. In other axiomatic systems, we will discover that the axioms actually lead to theorems telling us that there are many more points and/or lines than those guaranteed to exist by the axioms.

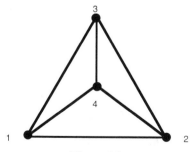

Figure 1.1

These observations suggest that the construction of any model for four-point geometry must begin with the objects known to exist, that is, four points. In model 1 these points are interpreted as the letters A, B, C, D whereas in model 2 (see Fig. 1.1) these points are interpreted as dots. In continuing to build either model, we must interpret the remaining undefined terms so as to create a system in which Axioms 2 and 3 become true statements.

Model 1

Undefined Term	Interpretation
Points	Letters A, B, C, D
Lines	Columns of letters given below
On	Contains or is contained in

$$
\begin{array}{cccccc}
A & A & A & B & B & C \\
B & C & D & C & D & D
\end{array}
$$

Model 2

Undefined Term	Interpretation
Points	Dots denoted 1, 2, 3, 4
Lines	Segments illustrated in Fig. 1.1
On	A dot is an endpoint of a segment or vice versa

There are several other important properties that an axiomatic system *may* possess.

Definition 1.2. An axiom in an axiomatic system is *independent* if it cannot be proved from the other axioms. If each axiom of a system is independent, the system is said to be independent.

Clearly an independent system is more elegant since no unnecessary assumptions are made. However, the increased difficulty of working in an independent system becomes obvious when we merely note that accepting fewer statements without proof leaves more statements to be proved. For this reason the axiomatic systems used in high-school geometry are seldom independent.

The verification that an axiomatic system is independent is also done via models. The independence of Axiom A in an axiomatic system S is established by finding a model of the system S' where S' is the system obtained from S by replacing Axiom A by a negation of A. Thus, to demonstrate that a system consisting of n axioms is independent, n models must be exhibited—one for each axiom. The independence of the axiomatic system for four-point geometry is demonstrated by the following three models, all of which interpret points as letters of the alphabet and lines as the columns of letters indicated.

Models Demonstrating Independence of Axioms for Four-Point Geometry

Model I1. A model in which a negation of Axiom 1 is true (i.e., there do not exist four points):

Points	Lines
A, B	A
	B

Since this model contains only two points, the negation of Axiom 1 is clearly true and it is easy to show that Axioms 2 and 3 are true statements in this interpretation.

Model I2. A model in which a negation of Axiom 2 is true (i.e., there are two distinct points not on one line):

Points	Lines	
A, B, C, D	A	C
	B	D

Note that in this model there is no line on points *A* and *C*. What other pairs of points fail to be on a line?

Model I3. A model in which a negation of Axiom 3 is true (i.e., there are lines not on exactly two points):

Points	Lines			
A, B, C, D	A	A	B	C
	B	D	D	D
	C			

In this model one line is on three points, whereas the remaining lines are each on two points, so the negation of Axiom 3 is true in this interpretation.

Since we have demonstrated the independence of each of the axioms of four-point geometry, we have shown that this axiomatic system is independent.

Another property that an axiomatic system may possess is completeness.

Definition 1.3. An axiomatic system is *complete* if every statement containing undefined and defined terms of the system can be proved valid or invalid, or in other words, if it is not possible to add a new independent axiom to the system.

In general, it is impossible to demonstrate directly that a system is complete. However, if a system is complete, there cannot exist two essentially different models. This means all models of the system must be pairwise *isomorphic*.

Definition 1.4. Two models α and β of an axiomatic system are said to be *isomorphic* if there exists a one-to-one correspondence ϕ from the set of points and lines of α onto the set of points and lines of β which preserves all relations. In particular if the undefined terms of the system consist of the terms "point," "line," and "incidence," then ϕ must satisfy the following conditions:

1. For each point P and line l in α, $\phi(P)$ and $\phi(l)$ are a point and line in β.
2. If P is incident with l, then $\phi(P)$ is incident with $\phi(l)$.

If all models of a system are pairwise isomorphic, it is clear that the models must each have the same number of points and lines. Furthermore, if a new independent axiom could be added to the system, there would be two distinct models of the system: a model α in which the new axiom would be valid and a model β in which the new axiom would *not* be valid. The models α and β could not then be isomorphic. Hence if all models of the system are necessarily isomorphic, it follows that the system is complete.

In the example of the four-point geometry, it is clear that models 1 and 2 are isomorphic. The verification that all models of this system are isomorphic follows readily once the following theorem is verified. (See Exercises 5 and 6.)

Theorem 1.1. *There are exactly six lines in the four-point geometry.*

Finally any discussion of the properties of axiomatic systems must include mention of the important result contained in Gödel's theorem. Greatly simplified, this result says that any consistent axiomatic system comprehensive enough to contain the results of elementary number theory is not complete.

EXERCISES

For Exercises 1–4, consider the following axiomatic system:

Axioms for Three-Point Geometry

Undefined Terms. Point, line, on.
Axiom 1. There exist exactly three points.
Axiom 2. Two distinct points are on exactly one line.
Axiom 3. Not all points are on the same line.
Axiom 4. Two distinct lines are on at least one common point.

1. (a) Prove that this system is consistent. (b) Did the proof in part (a) demonstrate absolute consistency or relative consistency? Explain.

2. Prove that this system is independent.

3. Prove the following theorems in this system: (a) Two distinct lines are on exactly one point. (b) Every line is on exactly two points. (c) There are exactly three lines.

4. Is this system complete? Why?

5. Prove Theorem 1.1.

6. Prove that any two models of four-point geometry are isomorphic.

Use the following definition in Exercises 7 and 8.

Definition. The *dual* of a statement p in the four-point geometry is obtained by replacing each occurrence of the term "point" in p by the term "line" and each occurrence of the term "line" in p by the term "point."

7. Obtain an axiomatic system for *four-line geometry* by dualizing the axioms for four-point geometry.

8. Verify that the dual of Theorem 1.1 will be a theorem of four-line geometry. How would its proof differ from the proof of Theorem 1.1 in Exercise 5?

1.3. Finite Projective Planes

As indicated by the examples in the previous section, there are geometries consisting of only a finite number of points and lines. In this section we will consider an axiomatic system for an important collection of finite geometries known as finite projective planes. These geometries may, at first glance, look much like finite versions of plane Euclidean geometry. However, there is a very important difference. In a finite projective plane, each pair of lines intersects; that is, there are no parallel lines. This pairwise intersection of lines leads to several other differences between projective planes and Euclidean planes. A few of these differences will become apparent in this section; others will not become evident until we study general plane projective geometry in Chapter 4.

Some of the first results in the study of finite projective geometries were obtained by von Staudt in 1856, but it wasn't until early in this century that finite geometries assumed a prominent role in mathematics. Since then, the study of these geometries has grown considerably and there are still a number of unsolved problems currently engaging researchers in this area.

Axioms for Finite Projective Planes

Undefined Terms. Point, line, incidence.
Defined Terms. Points incident with the same line are said to be *collinear*. Lines incident with the same point are said to be *concurrent*.

Axiom P1. There exist at least four distinct points, no three of which are collinear.

Axiom P2. There exists at least one line with exactly $n + 1$ $(n > 1)$ distinct points incident with it.

Axiom P3. Given two distinct points, there is exactly one line incident with both of them.

Axiom P4. Given two distinct lines there is at least one point incident with both of them.

Any set of points and lines satisfying these axioms is called a *projective plane of order n*. Note that the word "incidence" has been used as the third undefined term in this axiom system. The usage of this word rather than the word "on" is more common in the study of general projective planes.

The consistency of this axiomatic system is demonstrated by either of the following models which use the same interpretations as models 1 and 2 in Section 1.2.

Model 1

Points	Lines						
A, B, C, D, E, F, G	A	A	B	A	B	C	C
	B	D	D	F	E	D	E
	C	E	F	G	G	G	F

Model 2

Points	Lines
Dots denoted $1, 2, 3, 4, 5, 6, 7$	Segments illustrated in Fig. 1.2

Note that these models are projective planes of order two and both have exactly three points on each line, but there are models with more than three points on a line as shown by the next model.

 Model 3

Points	Lines											
$A, B, C, D, E,$	A	A	A	A	B	B	B	C	C	C	D	D	D
$F, G, H, I, J,$	B	E	H	K	E	F	G	E	F	G	E	F	G
K, L, M	C	F	I	L	H	I	J	I	J	H	J	H	I
	D	G	J	M	K	L	M	M	K	L	L	M	K

Whereas models 1 and 2 have three points on each line, three lines on each point, and a total of seven points and seven lines, model 3 has four points on each line, four lines on each point, and a total of 13 points and 13 lines. To

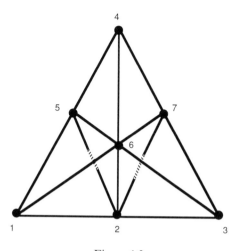

Figure 1.2

determine if finite projective planes exist with more points and lines, it is clearly impractical to employ trial-and-error procedures. Instead we develop a series of theorems that lead to a general result regarding the number of points and lines in a finite projective plane of order n.

The proofs of these theorems are simplified by noting that this axiom system satisfies the *principle of duality*, which Coxeter has described as "one of the most elegant properties of projective geometry" (Coxeter, 1969, p. 231). As noted in the exercises in Section 1.2, the *dual* of a statement is obtained by replacing each occurrence of the word "point" by the word "line" and vice versa (consequently, the words "concurrent" and "collinear" must also be interchanged).

Definition 1.5. An axiomatic system in which the dual of any theorem is also a theorem is said to satisfy the *principle of duality*.

Thus in an axiomatic system that satisfies the principle of duality, the proof of any theorem can be "turned into" a proof of a dual theorem merely by dualizing the original proof. To show that any axiom system has the property of duality it is necessary to prove that the duals of each axiom are theorems of the system. The theorems that are the dual statements of the four axioms of this system are listed here. The proofs of the duals of Axioms P1, P3, and P4 are left to you.

Theorem P1 (Dual of Axiom P1). *There exist at least four distinct lines, no three of which are concurrent.*

Theorem P2 (Dual of Axiom P3). *Given two distinct lines there is exactly one point incident with both.*

Theorem P3 (Dual of Axiom P4). *Given two distinct points, there is at least one line incident with both of them.*

Theorem P4 (Dual of Axiom P2). *There exists at least one point with exactly $n + 1$ ($n > 1$) distinct lines incident with it.*

Proof. By Axiom P2 there is a line l with $n + 1$ points $P_1, P_2, \ldots, P_{n+1}$ and by Axiom P1 there is a point P not incident with l. Then by Axiom P3 there exist lines $l_1, l_2, \ldots, l_{n+1}$ joining the point P to points $P_1, P_2, \ldots, P_{n+1}$, respectively (see Fig. 1.3). It is sufficient to show that these lines are all distinct and that there are no other lines through P. If $l_i = l_j$ for $i \neq j$ then the two points P_i and P_j would be incident with both l and $l_i = l_j$ and it would follow by Axiom P3 that $l = l_i = l_j$. But P is on l_i and *not* on l so we have a contradiction. Thus $l_i \neq l_j$ for $i \neq j$. Now assume there is an additional line, l_{n+2} through P. This line must also intersect l at a point Q (Axiom P4). Since l has exactly $n + 1$ points, Q must be one of the points $P_1, \ldots, P_{n+1}$. Assume $Q = P_1$, then since $Q = P_1$ and P are two distinct points on both l_1 and l_{n+2}, it follows that $l_{n+2} = l_1$. Therefore the point P is incident with exactly $n + 1$ lines. $\qquad\qquad\square$

The previous proof demonstrates several geometric conventions. First, to make the proof less awkward, the phrase "is incident with" is frequently replaced by a variety of other familiar terms such as "is on," "contains," and "through." The meanings of these substitute terms should be explained by their context. Second, capital letters are used to designate points while lowercase letters are used for lines. Finally, since diagrams are extremely helpful both in constructing and following a proof, figures are included as part of the proofs whenever appropriate; but the narrative portions of the proofs are constructed so as to be completely independent of the figures.

In models 1–3, the number of points on each line and the number of lines on each point is the same for all lines and points in each model. That this must be true in general is verified by the following theorems.

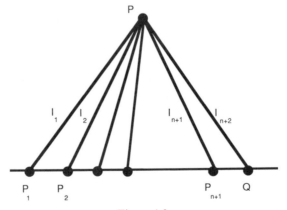

Figure 1.3

Theorem P5. *In a projective plane of order n, each point is incident with exactly $n + 1$ lines.*

Proof. Let P be a point of the plane. Axiom P2 guarantees the existence of a line l containing $n + 1$ points, $P_1, P_2, \ldots, P_{n+1}$. Then there are two cases to consider, depending on whether P is on l or not (see Figs. 1.4 and 1.5).

Case 1 (P is not on l). If P is not on l there are at least $n + 1$ lines through P, namely, the lines joining P to each of the points $P_1, P_2, \ldots, P_{n+1}$. Just as in the proof of the previous theorem, it can be shown that these lines are distinct and there are no other lines through P. So in this case there are exactly $n + 1$ lines through P.

Case 2 (P is on l). Assume $P = P_1$. Axiom P1 guarantees the existence of a point Q not on l. It is also possible to verify the existence of a line m, which contains neither P nor Q (see Exercise 7). By case 1, Q is on exactly $n + 1$ lines $m_1, m_2, \ldots, m_{n+1}$. But each of these lines intersects m in a point R_i for $i = 1, \ldots, n + 1$. It can easily be shown that these points are distinct and that these are the only points on line m. Thus P is not on the line m, which contains exactly $n + 1$ points, so as in case 1, P is incident with exactly $n + 1$ lines.   □

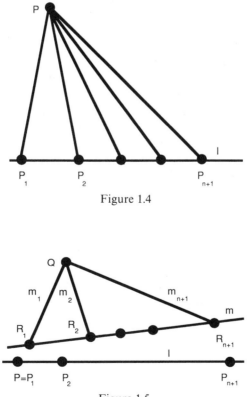

Figure 1.4

Figure 1.5

With this theorem in hand, the following theorem follows immediately by duality.

Theorem P6. *In a projective plane of order n, each line is incident with exactly $n + 1$ points.*

Using these results, we can now determine the total number of points and lines in a projective plane of order n.

Theorem P7. *A projective plane of order n contains exactly $n^2 + n + 1$ points and $n^2 + n + 1$ lines.*

Proof. Let P be a point in a projective plane of order n. Then every other point is on exactly one line joining it with the point P. By Theorem P5 there are exactly $n + 1$ lines through P and by Theorem P6 each of these lines contains exactly $n + 1$ points, that is, n points in addition to P. Thus the total number of points is $(n + 1)n + 1 = n^2 + n + 1$. A dual argument verifies that the total number of lines is also $n^2 + n + 1$.   ☐

Thus a finite projective plane of order two must have seven points and seven lines and a projective plane of order three must have 13 points and lines. But one of the unresolved questions in the study of finite geometries is the determination of the orders for which finite projective planes exist. A partial answer to this question was given in 1906 when Veblen and Bussey proved that there exist finite projective planes of order n, whenever n is a power of a prime. It has long been conjectured that these are the only orders for which finite projective planes exist. In 1949 Bruck and Ryser proved that if n is congruent to 1 or 2 (modulo 4) and if n cannot be written as the sum of two squares, then there are no projective planes of order n. This proved the conjecture for an infinite number of cases including $n = 6, 14, 21$, and 22. However, it also left open an infinite number of cases including $n = 10, 12, 15, 18$, and 20. In late 1988, a group of researchers in the computer science department at Concordia University in Montreal completed a case-by-case computer analysis requiring several thousand hours of computer time. By investigating the implications of the existence of an order 10 projective plane, they concluded that the conjecture is also correct for $n = 10$; that is, finite projective planes of order 10 do not exist. This leaves $n = 12$ as the smallest number for which the conjecture is unproved (Cipra, 1988).

The study of the infinite projective plane from both synthetic and analytic viewpoints yields a wealth of interesting geometric properties, which are generalizations of both Euclidean and non-Euclidean properties. We pursue this study in Chapter 4, following an introduction of non-Euclidean geometry (Chapter 2) and the development of an analytic model for Euclidean geometry (Chapter 3). However, as we shall see in the following section, even one of the simplest projective geometries, namely, the finite projective plane of order two, has an application that demonstrates the relevance of geometry to exciting new areas of mathematics.

EXERCISES

1. Which axioms for a finite projective plane are also valid in Euclidean geometry? Which are not?

2. Prove that the axiomatic system for finite projective planes is incomplete.

3. Verify that models 1 and 2 are isomorphic.

4. Prove Theorem P1.

5. Prove Theorem P2.

6. Prove Theorem P3.

7. Verify the existence of the line m used in case 2 of the proof of Theorem P5.

8. How many points and lines does a finite projective plane of order seven have?

The axioms for a *finite affine plane of order n* are given. The undefined terms and definitions are identical to those for a finite projective plane.

Axioms for Finite Affine Planes

Axiom A1. There exist at least four distinct points no three of which are collinear.
Axiom A2. There exists at least one line with exactly n ($n > 1$) points on it.
Axiom A3. Given two distinct points, there is exactly one line incident with both of them.
Axiom A4. Given a line l and a point P not on l, there is exactly one line through P that does not intersect l.

9. How do the axioms for a finite affine plane differ from those for a finite projective plane?

10. Show that a finite affine plane does not satisfy the principle of duality.

11. Find models of affine planes of orders two and three.

The following exercises ask you to prove a series of theorems about finite affine planes. You should prove these in the order indicated since some will require that you use a previous result.

12. Prove: In an affine plane of order n, each point lies on exactly $n + 1$ lines. [*Hint*: Consider two cases as in the proof of Theorem P5.]

13. Prove: In an affine plane of order n, each line contains exactly n points.

14. Prove: In an affine plane of order n, each line l has exactly $n - 1$ lines that do not intersect l.

15. Prove: In an affine plane of order n, there are exactly n^2 points and $n^2 + n$ lines.

16. Verify that if one line and its points are deleted from the finite projective plane of order 2 given in model 1 or 2, the remaining points and lines form a model of an affine plane. What is its order?

1.4. An Application to Error-Correcting Codes

The finite projective planes of order two demonstrated in models 1 and 2 of the previous section are known as *Fano planes*. A concise way of representing these and other finite planes is a configuration known as an *incidence table*. The lines of the plane are represented by columns in Table 1.1, while the points of the plane are represented by rows. Entries of 0 and 1 represent nonincidence and incidence, respectively.

This table demonstrates that we can represent each point in a Fano plane uniquely by a vector consisting of the entries in the corresponding row of the incidence table. Thus point A can be represented by the vector $(1, 0, 0, 0, 0, 1, 1)$. Similarly, every point in a Fano plane can be represented by a binary 7-tuple; that is, a vector with seven components, each of which is a 0 or 1. Note that the vector for any given point contains exactly three ones, so in the language of coding theory, we say that each vector has *weight* 3. Following a brief introduction to the area of coding theory, we shall see that these seven vectors play an important role in an elementary error-correcting code.

Coding theory is devoted to the detection and correction of errors that are introduced when messages are transmitted. Such codes have found application in sending pictures back from space and in the development of the compact disk. The impetus for developing these codes arose from the frustrations that Richard W. Hamming encountered in 1947 when working with a mechanical relay computer, which dumped his program whenever it detected an error.

Table 1.1. Incidence Table for a Fano Plane

	l_1	l_2	l_3	l_4	l_5	l_6	l_7
A	1	0	0	0	0	1	1
B	0	1	0	0	1	0	1
C	0	0	1	0	1	1	0
D	1	0	0	1	1	0	0
E	0	1	0	1	0	1	0
F	0	0	1	1	0	0	1
G	1	1	1	0	0	0	0

Since then, coding theory has become an important research area, using results from projective geometry, group theory, the theory of finite fields, and linear programming.

Hamming's initial frustration in having a computer that could detect but not find and correct an error led to the development of error-correcting codes. Error-correcting coding has been described as "the art of adding redundancy efficiently so that most messages, if distorted, can be correctly decoded" (Pless, 1982, p. 2).

One of the simplest error-correcting codes is a projective geometry code known as the Hamming (7, 4) code. This code can be generated by the four rows of the matrix G below. This matrix is known as the *generator matrix* for the code. In this matrix, the first row vector is the code word for 1000, the binary representation of the decimal number 8; the second row is the code word for 0100, the binary representation of the decimal number 4; and so on.

$$G = \begin{bmatrix} 1 & 0 & 0 & 0 & 0 & 1 & 1 \\ 0 & 1 & 0 & 0 & 1 & 0 & 1 \\ 0 & 0 & 1 & 0 & 1 & 1 & 0 \\ 0 & 0 & 0 & 1 & 1 & 1 & 1 \end{bmatrix}$$

Other code words are obtained by adding these rows where the addition is the usual componentwise vector addition modulo 2. Note that when we find all possible sums of these rows (see Table 1.2), we obtain in the first four positions all 16 possible strings of 0's and 1's; that is, all binary representations of the decimal numbers 0 through 15.

Table 1.2. Possible Code Words

0 0 0 0	0 0 0	Adding no words
1 0 0 0	0 1 1	Adding one word
0 1 0 0	1 0 1	
0 0 1 0	1 1 0	
0 0 0 1	1 1 1	
1 1 0 0	1 1 0	Adding two words
1 0 1 0	1 0 1	
1 0 0 1	1 0 0	
0 1 1 0	0 1 1	
0 1 0 1	0 1 0	
0 0 1 1	0 0 1	
1 1 1 0	0 0 0	Adding three words
1 1 0 1	0 0 1	
1 0 1 1	0 1 0	
0 1 1 1	1 0 0	
1 1 1 1	1 1 1	Adding four words

The first four digits of these code words occupy the so-called *information positions*, since they represent the actual number or message to be transmitted. The remaining three positions are called the *redundancy positions*. The digits in these last positions allow single error corrections; that is, if a transmitted message contains a single digit error these extra digits allow us to find and correct the error. For example, the message $x = 1010010$ does not appear in Table 1.2 as a possible code word. Assuming that a single error has occurred in the transmission of a code word we can locate the error and correct it using the parity check matrix, H. This parity check matrix consists of seven column vectors, which give the binary representations of the decimal numbers 1 through 7.

$$Hx = \begin{bmatrix} 0 & 0 & 0 & 1 & 1 & 1 & 1 \\ 0 & 1 & 1 & 0 & 0 & 1 & 1 \\ 1 & 0 & 1 & 0 & 1 & 0 & 1 \end{bmatrix} \begin{bmatrix} 1 \\ 0 \\ 1 \\ 0 \\ 0 \\ 1 \\ 0 \end{bmatrix} = \begin{bmatrix} 1 \\ 0 \\ 0 \end{bmatrix}$$

Since the result is $(1, 0, 0)$, namely, the binary representation of the decimal number 4, the error occurs in the fourth position; hence the original code word was 1011010. Similarly, we can show that each of the 2^7 possible binary 7-tuples differs from a possible code word in at most one digit; and if there is a difference, the digit in which the "error" occurs can be located with the parity check matrix. However, when an actual code word is multiplied by this parity check matrix, the result is $(0, 0, 0)$ (see Exercises 6 and 7).

This parity check matrix, H, can be thought of as the defining matrix for this code. Note that the matrix H clearly has rank 3, and since H is a 3×7 matrix it represents a linear transformation from a vector space of dimension 7 to one of dimension 3. As we recall from linear algebra the kernel of this linear transformation is the set of solutions of $Hx = 0$ and the dimension of this kernel is $7 - 3 = 4$. By demonstrating that $Hx = 0$ whenever x is a code word, we can show that the row vectors of the generator matrix G are basis vectors for this kernel. Thus the code words of the Hamming $(7, 4)$ code form a subspace of a vector space. Any code for which the code words form a subspace is said to be *linear*.

The code words of the Hamming $(7, 4)$ code can be considered to be coordinates of points in a seven-dimensional space where the entire space consists of points corresponding to the 2^7 possible messages, that is, the possible binary 7-tuples. Distance in this space is defined in terms of a function known as the *Hamming distance*.

Definition. The *Hamming distance* between two binary n-tuples x and y, $d(x, y)$, is given by the number of components in which the n-tuples differ.

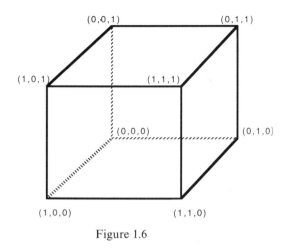

Figure 1.6

Thus if $x = 1001110$ and $y = 1011101$, $d(x, y) = 3$. Clearly the maximum distance between binary 7-tuples is 7 and as you can easily verify the minimum distance between any pair of nonzero code words in the Hamming $(7, 4)$ code is 3. Since this minimum distance is 3 this is also known as the Hamming $(7, 4, 3)$ code. Also note that the distance between 0000000 and any other binary 7-tuple x is just the number of ones in x, that is, the *weight* of x. Thus the *minimum weight* of this code is said to be 3.

Using this distance we can view the code words under consideration as select vertices in a seven-dimensional cube where edges join pairs of vertices with coordinates that differ in exactly one component, that is, pairs of vertices whose Hamming distance is 1. A diagram illustrating a three-dimensional cube is shown in Fig. 1.6. As you can see from this figure, the Hamming distance between two vertices in the three-dimensional cube counts the number of edges of the cube that must be traversed to get from one vertex to the other.

To illustrate the role of the Hamming distance in the Hamming $(7, 4, 3)$ code, we first consider a more elementary code consisting of the code words 000 and 111, which can be modeled by the three-dimensional cube. Clearly, the Hamming distance between these two code words is 3. Furthermore, the messages that could occur if exactly one error is made in transmitting the code word 000 would be 001, 010, and 100. These are represented by vertices at a Hamming distance two from the vertex representing the other possible code word, 111. The set of code words, $\{001, 010, 100\}$ can be thought of as a set of points at a distance 1 from the point 000. Thus this set is said to form a *1-sphere*, centered at the code word 000. Similarly, the remaining three possible error messages form a 1-sphere centered at the code word 111 (see Fig. 1.7). These two spheres partition the set of binary 3-tuples, so that every possible binary 3-tuple appears in exactly one 1-sphere. Thus if we assume that a given

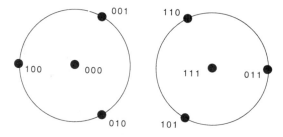

Figure 1.7. Spheres of binary 3-tuples.

message contains one or fewer errors, we can decode it by locating the unique nearest possible code word.

Similarly in the Hamming $(7, 4, 3)$ code, the minimum distance between code words is also 3 and all possible binary 7-tuples lie in a set of nonoverlapping 1-spheres, which exhaust the seven-dimensional cube (see Exercise 6). The decoding process we are using is that of locating the nearest possible code word. Codes that have this property that all possible messages lie within or on nonoverlapping spheres of radius t are called *perfect t-error-correcting codes*. Thus the Hamming $(7, 4, 3)$ code is a perfect 1-error-correcting code.

A result from coding theory (Blake, 1975, p. 185) shows that a perfect linear code is spanned by its minimum weight vectors. This means that the vectors of weight 3 span the Hamming $(7, 4, 3)$ code. As we can easily verify, these are the vectors in the rows of the incidence table for the Fano plane. Furthermore, the rows of the generator matrix G form a basis for this set.

EXERCISES

1. Show that the points and lines of the incidence table (Table 1.1) satisfy the axioms for a projective plane.

2. Demonstrate that the Fano plane given by the incidence table (Table 1.1) is isomorphic to that given in model 1 of Section 1.3.

3. Verify that any pair of coordinate vectors in the incidence table (Table 1.1) differ in exactly four components, that is, their Hamming distance is 4.

4. Write out the binary representations of the decimal numbers 1 through 15.

5. Verify that there are exactly 2^7-16 binary 7-tuples that are not code words in the Hamming $(7, 4, 3)$ code.

6. (a) Show that there are exactly seven binary 7-tuples that differ from the code word 1000011 in exactly one digit. (b) Apply the parity check matrix H to one of these seven and verify that it does locate the position in which the digit differs.

7. Show that $Hx = 0$ for each row vector in the generator matrix G.

8. Obtain all possible code words in the linear $(5, 3)$ binary code with generator matrix G'.

$$G' = \begin{bmatrix} 1 & 0 & 0 & 1 & 1 \\ 0 & 1 & 0 & 0 & 1 \\ 0 & 0 & 1 & 1 & 1 \end{bmatrix}$$

9. Show that the Hamming distance is a metric, that is, that it satisfies each of the following conditions:

   (i) $d(x, y) = 0$ iff $x = y$.
   (ii) $d(x, y) = d(y, x)$.
   (iii) $d(x, z) \leq d(x, y) + d(y, z)$.

10. Verify that the minimum distance between any pair of code words in the Hamming $(7, 4, 3)$ code is 3.

11. Show that the set of code words in the Hamming $(7, 4, 3)$ code can also be obtained by adding the two 7-tuples 0000000 and 1111111 to the fourteen 7-tuples that occur either as rows of Table 1.1 or as rows of the incidence table obtained from Table 1.1 by interchanging 0's and 1's.

1.5. Desargues' Configurations

In this section we consider an axiomatic system for one more finite structure. We shall see that this structure not only satisfies the principle of duality but also exhibits an interesting relation between points and lines similar to the *polarity* relation of projective geometry. This relation involves points that do not lie on a line. Since the term "geometry" is usually reserved for structures in which each pair of points determines a unique line, we refer to the structures that satisfy our axioms as Desargues' configurations. Desargues' configurations are so named because they illustrate a theorem in real projective geometry known as Desargues' theorem. This theorem is stated in terms of two particular properties of triangles, that is, sets of three noncollinear points. If two triangles ABC and DEF have the property that lines joining corresponding vertices (i.e., AD, BE, CF) are concurrent, the triangles are said to be *perspective from a point*. Similarly, if the triangles possess the dual property that the intersections of corresponding sides are collinear they are said to be *perspective from a line*. With these definitions, Desargues' theorem can be stated succinctly.

Desargues' Theorem. *If two triangles are perspective from a point, then they are perspective from a line.*

An example of a Desargues' configuration and its corresponding incidence

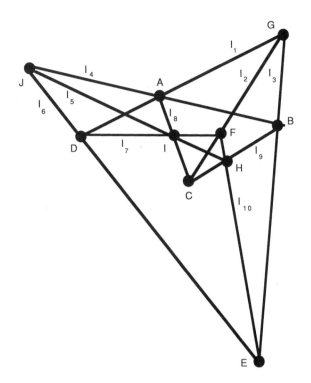

Figure 1.8. A Desargues' configuration.

table is shown in Fig. 1.8 and Table 1.3 (as in Section 1.4 entries of 0 and 1 represent nonincidence and incidence, respectively). As you can see either from the configuration or the incidence table, ABC and DEF are triangles that are perspective from point G and line l_5.

Careful scrutiny of either the structure shown in Fig. 1.8 or the corresponding

Table 1.3. Incidence Table for a Desargues' Configuration

	l_1	l_2	l_3	l_4	l_5	l_6	l_7	l_8	l_9	l_{10}
A	1	0	0	1	0	0	0	1	0	0
B	0	0	1	1	0	0	0	0	1	0
C	0	1	0	0	0	0	0	1	1	0
D	1	0	0	0	0	1	1	0	0	0
E	0	0	1	0	0	1	0	0	0	1
F	0	1	0	0	0	0	1	0	0	1
G	1	1	1	0	0	0	0	0	0	0
H	0	0	0	0	1	0	0	0	1	1
I	0	0	0	0	1	0	1	1	0	0
J	0	0	0	1	1	1	0	0	0	0

incidence table (Table 1.3) will lead to the observation that for each point M in the structure there is a line m such that no lines join M with points on m. The point M and line m are referred to as *pole* and *polar*, respectively. This *pole–polar* relation is described in detail by the following definitions and axioms.

Axioms for Desargues' Configurations

Undefined Terms. Point, line, on.
Defined Terms. If there are no lines joining a point M with points on line m (M not on m), m is called a *polar* of M and M is called a *pole* of m.
Axiom D1. There exists at least one point.
Axiom D2. Each point has at least one polar.
Axiom D3. Every line has at most one pole.
Axiom D4. Two distinct points are on at most one line.
Axiom D5. There are exactly three distinct points on every line.
Axiom D6. If line m does not contain point P, then there is a point on both m and any polar of P.

It should be no surprise that the Desargues' configuration shown in Fig. 1.8 provides a model for this axiomatic system. Furthermore, as you can easily verify, this axiomatic system satisfies the principle of duality (see Exercise 3).

Other properties of Desargues' configurations are given by the following theorems. The first of these theorems describes an important property of poles and polars. We encounter this property again when we study the polarity relation in projective geometry in Chapter 4.

Theorem D1. *If P is on a polar of point Q, then Q is on each polar of P.*

Proof. Let P be on q where q is a polar of Q (see Fig. 1.9). Thus since Q is not on q (why?), q must contain two more points R and S, which are distinct from P and Q (Axiom D5). Let p be a polar of P and assume Q is not on p. Then by Axiom D6, p and q must intersect at a point, namely, P, R, or S. But P is not on p by definition. And if R or S are on p, then q is a line joining P with a point on its polar, contradicting the definition. Thus Q is on p.   □

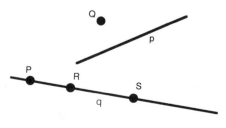

Figure 1.9

The usefulness of the property described in Theorem D1 is illustrated in the proofs of the following two theorems, which verify that the correspondence between poles and polars is one-to-one.

Theorem D2. *Every point has exactly one polar.*

Proof. Let P be an arbitrary point. By Axiom D2, P has at least one polar p. Assume P has a second polar p'. By Axioms D4 and D5 there is a point T on p' but not on p. Let t be a polar of T. Then by Axiom D6, p and t intersect. But since T is on p', P is on t by the previous theorem, and so line t joins P to a point on p, contradicting the definition of polar. Thus P has exactly one polar. $\square$

Theorem D3. *Every line has exactly one pole.*

Proof. By Axiom D3 every line has at most one pole. Hence it suffices to show that an arbitrary line p has at least one pole. Let R, S, and T be the three points on p, and let r and s be the unique polars of R and S (Theorem D2). Clearly, S is not on r (why?). Therefore, by Axiom D6, there is a point P on r and s. But P is on the polars of R and S; hence by Theorem D1, R and S are on the unique polar of P. So p is the polar of P or P is the pole of p. $\square$

These theorems and the following exercises illustrate that even though a finite structure may involve a limited number of points and lines, the structure may possess "strange" properties such as duality and polarity, which are not valid in Euclidean geometry. Another unexpected property illustrated by the exercises is that in Desargues' configurations a line has exactly three lines parallel to it through its pole; that is, there are points through which there are three lines parallel to a given line (see Exercise 6). Because of this latter property, Desargues' configurations can be classified as non-Euclidean.

EXERCISES

1. In the Desargues' configuration shown in Fig. 1.8 find the pole of line AB and the polar of C.

2. (a) Find two triangles in the Desargues' configuration in Fig. 1.8 that are perspective from point C. From which line are these two triangles perspective? (b) Find two triangles in the Desargues' configuration in Fig. 1.8 that are perspective from line AB. From which point are these two triangles perspective?

The following exercises ask you to verify theorems in the axiom system for Desargues' configurations. This means you must justify your proofs on the basis of the axioms—you cannot verify your reasoning on the basis of the model or incidence table given in this section.

3. Verify the duals of Axioms D1–D6.

4. Prove: There is a line through two distinct points iff their polars intersect.

5. Prove: If p and q are two lines both parallel to m (i.e., p and m have no common points, nor do q and m), then p and q intersect at the pole of m.

6. Prove: Through a point P there are exactly three lines parallel to p, the polar of P (i.e., the three lines have no points in common with line p).

7. Prove: There are exactly 10 points and 10 lines in a Desargues' configuration.

8. Prove Desargues' theorem. That is, show that if ABC and $A'B'C'$ are two triangles perspective from a point P, then they are perspective from a line. (Assume that the points A, B, C, A', B', C' and P are all distinct and that no three of the points A, B, C, A', B', C' are collinear.)

The following exercises ask you to work in an axiomatic system for finite structures known as *Pappus' configurations*. These axioms are as follows:

Axioms for Pappus' Configurations

Undefined Terms. Point, line, on.
Defined Terms. Two lines without a common point on them are *parallel*. Two points without a common line on them are *parallel*.
Axiom P1. There exists at least one line.
Axiom P2. There are exactly three distinct points on every line.
Axiom P3. Not all points are on the same line.
Axiom P4. There is at most one line on any two distinct points.
Axiom P5. If P is a point not on a line m, there is exactly one line on P parallel to m.
Axiom P6. If m is a line not on a point P, there is exactly one point on m parallel to P.

9. (a) Construct a model of a Pappus' configuration. (b) Construct an incidence table for this model.

10. Verify that this axiomatic system satisfies the principle of duality.

11. Prove: If m is a line, there are exactly two lines parallel to m.

12. Prove: There are exactly nine points and nine lines in a Pappus' configuration.

13. Prove: If m and n are parallel lines with distinct points A, B, C on m and A', B', C' on n, then the three points of intersections of AC' and CA', AB' and BA', BC' and CB' are collinear. (This result, which is valid in some projective planes, is known as the Theorem of Pappus.)

1.6. Suggestions for Further Reading

Albert, A.A., and Sandler, R. (1968). *An Introduction to Finite Projective Planes.* New York: Holt, Rinehart and Winston. (Contains a thorough group theoretic treatment of finite projective planes.)

Anderson, I. (1974). *A First Course in Combinatorial Mathematics.* Oxford, England: Clarendon Press. (Chapter 6 discusses block designs and error-correcting codes.)

Beck, A., Bleicher, M.N., and Crowe, D.W. (1972). *Excursions into Mathematics.* New York: Worth. (Sections 4.9–4.15 give a very readable discussion of finite planes, including the development of analytic models.)

Benedicty, M., and Sledge, F.R. (1987). *Discrete Mathematical Structures.* Orlando, FL: Harcourt Brace Jovanovich. (Chapter 13 gives an elementary presentation of coding theory.)

Gensler, H.J. (1984). *Gödel's Theorem Simplified.* Lanham, MD: University Press of America.

Hofstadter, D.R. (1984). Analogies and metaphors to explain Gödel's theorem. In: D.M. Campbell and J.C. Higgins (Eds.), *Mathematics: People, Problems, Results,* Vol. 2, pp. 262–275. Belmont, CA: Wadsworth.

Kolata, G. (1982). Does Gödel's theorem matter to mathematics? *Science* 218: 779–780.

Lam, C.W.H. (1991). The Search for a Projective Plane of Order 10. *The American Mathematical Monthly.* Vol. 98, No. 4, pp. 305–318.

Lockwood, J.R., and Runion, G.E. (1978). *Deductive Systems: Finite and non-Euclidean Geometries.* Reston, VA: N.C.T.M. (Chapter 1 contains an elementary discussion of axiomatic systems.)

Nagel, E., and Newman, J.R. (1956). Gödel's proof. In: J.R. Newman (Ed.), *The World of Mathematics,* Vol. 3, pp. 1668–1695. New York: Simon and Schuster.

Pless, V. (1982). *Introduction to the Theory of Error-Correcting Codes.* New York: Wiley. (A well-written explanation of this new discipline and the mathematics involved.)

Smart, J.R. (1978). *Modern Geometries,* 2nd ed. Belmont, CA: Wadsworth. (Chapter 1 contains an easily readable discussion of axiomatic systems and several finite geometries.)

Thompson, T.M. (1983). *From Error-Correcting Codes Through Sphere Packings to Simple Groups.* The Carus Mathematical Monographs, No. 21. Ithaca, NY: M.A.A. (Incorporates numerous historical antecdotes while tracing 20th century mathematical developments involved in these topics.)

Readings on Latin Squares

Beck, A., Bleicher, M.N., and Crowe, D.W. (1972). *Excursions into Mathematics,* pp. 262–279. New York: Worth.

Crowe, D.W., and Thompson, T.M. (1987). Some modern uses of geometry. In: M.M. Lindquist and A.P. Schulte (Eds.). *Learning and Teaching Geometry, K–12,* 1987 Yearbook, pp. 101–112. Reston, VA: N.C.T.M.

Gardner, M. (1959). Euler's spoilers: The discovery of an order-10 Graeco-Latin square. *Scientific American* 201: 181–188.

Sawyer, W.W. (1971). Finite arithmetics and geometries. In: *Prelude to Mathematics,* Chap. 13. New York: Penguin Books.

CHAPTER 2

Non-Euclidean Geometry

2.1. Gaining Perspective

Mathematics is not usually considered a source of surprises, but non-Euclidean geometry contains a number of easily obtainable theorems that seem almost "heretical" to anyone grounded in Euclidean geometry. A brief encounter with these "strange" geometries frequently results in initial confusion. Eventually, however, this encounter should not only produce a deeper understanding of Euclidean geometry but also offer convincing support for the necessity of carefully reasoned proofs for results that may have once seemed obvious. These individual experiences mirror the difficulties mathematicians encountered historically in the development of non-Euclidean geometry. An acquaintance with this history and an appreciation for the mathematical and intellectual importance of Euclidean geometry is essential for an understanding of the profound impact of this development on mathematical and philosophical thought. Thus, the study of Euclidean and non-Euclidean geometry as mathematical systems can be greatly enhanced by parallel readings in the history of geometry. Since the mathematics of the ancient Greeks was primarily geometry, such readings provide an introduction to the history of mathematics in general.

The sources recommended at the end of this chapter are intended to provide insight into the following:

1. The nature and uses of geometry in ancient civilizations like those of Babylon, China, and Egypt.
2. The mystical qualities that were associated with mathematical and geometric relations by groups like the Pythagoreans.
3. The importance of compass and straight edge constructions and the dilemma posed by three particular construction problems.
4. The emergence of deductive reasoning and the understanding of the nature of axioms in ancient Greece.
5. The importance of Euclidean geometry in the philosophies of Plato and Kant.

6. The reasons for the repeated attempts to prove Euclid's fifth postulate.
7. The numerous "false starts" and the reasons for the delay in the development of non-Euclidean geometry.
8. The influence of the development of non-Euclidean geometry on mathematical and philosophical thought.

2.2. Euclid's Geometry

To understand the significance of non-Euclidean geometry it is essential to become familiar with the geometry developed by the ancient Greeks. This geometry reached its apogee in approximately 300 B.C., with the appearance of the *Elements* of Euclid. In the 13 books comprising this treatise, Euclid organized 465 propositions summarizing the currently known results not only in geometry but also in number theory and elementary (geometric) algebra.

The *Elements* of Euclid is important for its significant mathematical content, but it also has become a landmark in the history of mathematics, because it is the earliest extensive example of the use of the axiomatic method. Euclid realized that not every mathematical statement can be proved, that certain statements must be accepted as basic assumptions. Euclid referred to these assumptions as postulates and common notions, but they are now known as axioms.

Euclid's work was immediately accorded the highest respect and recognized as a work of genius. As a result all previous work in geometry was quickly overshadowed so that now there exists little information about earlier efforts. It is a further mark of the monumental importance of this work, that the *Elements* was used essentially unmodified as a standard geometry text for centuries. (The geometry contained in the *Elements* became known as Euclidean geometry.)

Euclid's definitions, postulates, common notions, and first 30 propositions as translated by Sir Thomas Heath are given in Appendix A. A careful consideration of these should make the following observations apparent.

1. Even though Euclid realized the necessity of axioms, he apparently did not realize the need for undefined terms. However, a consideration of the first seven definitions suggests that these seven terms are essentially undefined.
2. In the listing of his axioms, Euclid makes a distinction between those he called postulates and those he called common notions. The former were supposedly geometric in nature, whereas the latter were supposedly common to all mathematics.
3. The statement of the fifth postulate is much more involved than the other four.

This third observation led geometers to suspect that this statement was not independent of the first four postulates; but that it could be proved on the basis of the common notions and the first four postulates. The fact that Euclid had proved his first 28 propositions (theorems) without resorting to the use of this postulate added fuel to this speculation. The attempts to prove the fifth postulate began soon after the appearance of the *Elements*. In these attempts, geometers frequently made an assumption and used this assumption to prove the fifth postulate. However, each of these assumptions was eventually proved equivalent to the fifth postulate (in the presence of other axioms sufficient to recapture Euclid). A list of some of these equivalent formulations is both interesting and instructive. One of these statements involves the notion of equidistant lines. Recall that the *distance from a point P to a line m* is the length of the perpendicular segment from P to m. If the distance from each point on a line l to line m is the same, then l is said to be *equidistant* from m.

Statements Equivalent to Euclid's Fifth Postulate:

1. *Playfair's Axiom.* Through a given point not on a given line exactly one parallel can be drawn to a given line.
2. The sum of the angles of any triangle is equal to two right angles.
3. There exists a pair of similar triangles.
4. There exists a pair of straight lines everywhere equidistant from one another.
5. Given any three noncollinear points, there exists a circle passing through them.
6. If three angles of a quadrilateral are right angles, then the fourth angle is also a right angle.

The proof of the equivalence of the fifth postulate and Playfair's axiom is presented later. This proof demonstrates the two steps required to prove that a statement is equivalent to Euclid's fifth postulate: (1) We must construct a proof of Playfair's axiom using Euclid's five postulates; and (2) we must construct a proof of Euclid's fifth postulate using Euclid's first four postulates together with Playfair's axiom. Note that in both (1) and (2), Euclid's Propositions 1–28 can be used. Proofs of the equivalence of the fifth postulate and statements 2–6 are contained in *Introduction to non-Euclidean Geometry* by Harold E. Wolfe (1945).

Proof of the Equivalence of Playfair's Axiom and Euclid's Fifth Postulate

Proof of Playfair's Axiom Based on Euclid's Postulates. Let the given point be P and the line be l (see Fig. 2.1). Through P construct a line perpendicular to l at Q (Proposition 12). Then through P construct a second line PR perpendicular to PQ (11). The lines PR and l are parallel (27). Now assume PS is a

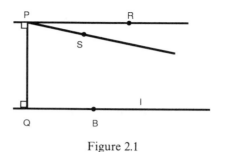

Figure 2.1

second line through P as shown. Then $\angle QPS$ is less than $\angle QPR$ (C.N. 5). Hence $\angle BQP$ and $\angle QPS$ (where B is a point on l as shown) are together less than $\angle BQP$ and $\angle QPR$ (C.N. 1). But $\angle BQP$ and $\angle QPR$ are right angles; therefore PS and l are not parallel by postulate 5.   □

Proof of the Fifth Postulate Based on Postulates 1–4 and Playfair's Axiom. Let AB and CD be lines cut by a transversal PQ so that $\angle DQP$ and $\angle QPB$ are together less than two right angles. At P, construct line PE so that $\angle DQP$ and $\angle QPE$ are together equal to two right angles (Proposition 23). Then PE is parallel to QD (28). So by Playfair's axiom, AB is not parallel to CD and thus AB and CD intersect (see Fig. 2.2). Now assume AB and CD intersect in a point S on the other side of PQ (see Fig. 2.3). Then $\angle SPQ$ and $\angle SQP$ are together greater than two right angles. But this contradicts Proposition 17. So AB and CD intersect on the appropriate side.   □

Returning to a consideration of Euclid's work, it is useful to investigate some of the proofs presented in the *The Thirteen Books of Euclid's Elements* as translated by Heath. In particular, we consider the proofs of Propositions 1, 16, 21, and 27, reprinted with the permission of Cambridge University Press. These proofs demonstrate some of the geometric properties that Euclid took for granted, that is, properties he assumed without stating them explicitly as postulates or common notions. The essential role of these properties in

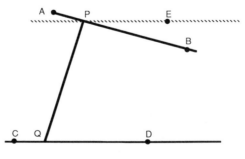

Figure 2.2

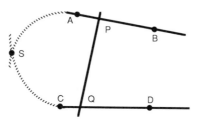

Figure 2.3

Euclid's geometry became evident with the development of non-Euclidean geometries.

Euclid's Proposition 1

On a given finite straight line to construct an equilateral triangle.
Let *AB* be the given finite straight line.
Thus it is required to construct an equilateral triangle on the straight line *AB*.
With centre *A* and distance *AB* let the circle *BCD* be described;

[Post. 3]

again, with centre *B* and distance *BA* let the circle *ACE* be described;

[Post. 3]

and from the point *C*, in which the circles cut one another, to the points *A*, *B* let the straight lines *CA*, *CB* be joined.   [Post. 1]

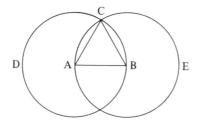

Now, since the point *A* is the centre of the circle *CDB*,

AC is equal to AB.   [Def. 15]

Again, since the point *B* is the centre of the circle *CAE*,

BC is equal to BA.   [Def. 15]

But *CA* was also proved equal to *AB*; therefore each of the straight lines *CA*, *CB* is equal to *AB*.
And things which are equal to the same thing are also equal to one another;   [C.N. 1]

therefore *CA* is also equal to *CB*.

Therefore the three straight lines *CA, AB, BC* are equal to one another.
Therefore the triangle *ABC* is equilateral; and it has been constructed on the given finite straight line *AB*.   (Being) what it was required to do.

In the proof of Proposition 1 Euclid assumed that the circles in his construction intersect in a point *C*; that is, he assumed the *continuity* of circles, without previously proving this as a proposition or stating it as a postulate. Later axiom systems for Euclidean geometry included explicit axioms of continuity, for example, Dedekind's axiom of continuity (see Appendix B). Note that Dedekind's axiom requires the concept of betweenness, which was also assumed by Euclid.

Euclid's Proposition 16

In any triangle, if one of the sides be produced, the exterior angle is greater than either of the interior and opposite angles.
Let *ABC* be a triangle, and let one side of it *BC* be produced to *D*;
I say that the exterior angle *ACD* is greater than either of the interior and opposite angles *CBA, BAC*.
Let *AC* be bisected at *E*   [1.10],
and let *BE* be joined and produced in a straight line to *F*;
let *EF* be made equal to *BE*   [1.3],
let *FC* be joined   [Post. 1],
and let *AC* be drawn through to *G*   [Post. 2].

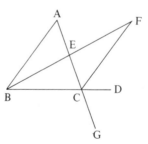

Then, since *AE* is equal to *EC*, and *BE* to *EF*,
the two sides *AE, EB* are equal to the two sides *CE, EF* respectively;
and the angle *AEB* is equal to the angle *FEC*, for they are vertical angles.   [1.15]
Therefore the base *AB* is equal to the base *FC*, and the triangle *ABE* is equal to the triangle *CFE*, and the remaining angles are equal to the remaining angles respectively, namely those which the equal sides subtend;   [1.4]
therefore the angle *BAE* is equal to the angle *ECF*.
But the angle *ECD* is greater than the angle *ECF*;   [C.N. 5]

therefore the angle *ACD* is greater than the angle *BAE*.

Similarly also, if *BC* be bisected, the angle *BCG*, that is, the angle *ACD* [1.15], can be proved greater than the angle *ABC* as well.

Therefore etc.   Q.E.D.

In Proposition 16 Euclid extended segment *BE* to a segment twice as long (*BF*). In doing so he implicitly assumed that a segment can be extended without coming back on itself so that *F* does not lie on segment *BE*. In more formal language, Euclid assumed that a line is infinite in extent, not merely boundless. As an example of a line that is boundless, but not of infinite extent, interpret the undefined term "line" as a great circle on a sphere.

Euclid's Proposition 21

If on one of the sides of a triangle, from its extremities, there be constructed two straight lines meeting within the triangle, the straight lines so constructed will be less than the remaining two sides of the triangle, but will contain a greater angle.

On *BC*, one of the sides of the triangle *ABC*, from its extremities *B*, *C*, let the two straight lines *BD*, *DC* be constructed meeting within the triangle;

I say that *BD*, *DC* are less than the remaining two sides of the triangle *BA*, *AC*, but contain an angle *BDC* greater than the angle *BAC*.

For let *BD* be drawn through to *E*.

Then, since in any triangle two sides are greater than the remaining one,   [1.20]

therefore, in the triangle *ABE*, the two sides *AB*, *AE* are greater than *BE*.

Let *EC* be added to each;

therefore *BA*, *AC* are greater than *BE*, *EC*.

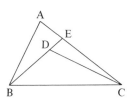

Again, since, in the triangle *CED*,

the two sides *CE*, *ED* are greater than *CD*,

let *DB* be added to each;

therefore *CE*, *EB* are greater than *CD*, *DB*.

But *BA*, *AC* were proved greater than *BE*, *EC*;

therefore *BA*, *AC* are much greater than *BD*, *DC*.

Again, since in any triangle the exterior angle is greater than the interior and opposite angle,   [1.16]
therefore, in the triangle *CDE*,

the exterior angle *BDC* is greater than the angle *CED*.

For the same reason, moreover, in the triangle *ABE* also,

the exterior angle *CEB* is greater than the angle *BAC*. But the angle *BDC* was proved greater than the angle *CEB*;

therefore the angle *BDC* is much greater than the angle *BAC*.

Therefore etc.   Q.E.D.

In Proposition 21, Euclid assumed that a line that contains the vertex (*B*) of a triangle ($\triangle ABC$) and an interior point (*D*) must intersect the opposite side (*AC*) at a point (*E*). Either this assumption or its equivalent, which was formulated by Pasch in the 19th century, is known as Pasch's axiom. The equivalent forms are as follows.

Pasch's Axiom (1). A line containing a vertex of a triangle and a point interior to the triangle will intersect the opposite side of the triangle.

Pasch's Axiom (2). Let A, B, C be three points not on the same line and let l be a line in the plane containing A, B, C that does not pass through $A, B,$ or C. Then if l passes through a point of the segment AB and contains a point interior to $\triangle ABC$, it will also pass through a point of segment AC or a point of segment BC.

Pasch's axiom can be shown to be equivalent to the separation properties of points and lines that Euclid also assumed; that is, he assumed that a point separates a segment into two distinct sets and that a line separates the plane into two distinct sets.

Euclid's Proposition 27

If a straight line falling on two straight lines make the alternate angles equal to one another, the straight lines will be parallel to one another.

For let the straight line *EF* falling on the two straight lines *AB, CD* make the alternate angles *AEF, EFD* equal to one another;

I say that *AB* is parallel to *CD*.

For, if not, *AB, CD* when produced will meet either in the direction of *B, D* or towards *A, C*.

Let them be produced and meet, in the direction of *B, D*, at *G*.

Then, in the triangle *GEF*, the exterior angle *AEF* is equal to the interior and opposite angle *EFG*:

which is impossible.   [1.16]

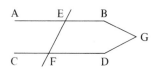

Therefore AB, CD when produced will not meet in the direction of B, D.

Similarly it can be proved that neither will they meet towards A, C.

But straight lines which do not meet in either direction are parallel;

[Def. 23]

therefore AB is parallel to CD.

Therefore etc.   Q.E.D.

The major point to be made in considering Proposition 27 is that the proof of this proposition, which guarantees the existence of parallel lines depends on the validity of Proposition 16, whose proof in turn requires that lines be of infinite extent.

These "shortcomings" in Euclid's work did not become significant until the development of non-Euclidean geometry. But then they presented a very real dilemma and had to be resolved. As a result, a number of new axiom systems for Euclidean geometry were developed. Obviously these systems were necessarily much longer and more involved than Euclid's. Appendixes B, C, and D contain systems developed by Hilbert, Birkhoff, and School Mathematics Study Group (S.M.S.G.), respectively. Examine these to determine how each system eliminates the shortcomings encountered in Euclid. Note that Appendix F contains proofs of one theorem (the angle-side-angle theorem) in all three systems.

EXERCISES

1. Prove the following statements equivalent to Euclid's fifth postulate: (a) If a line intersects one of two parallel lines, it also intersects the other. (b) Straight lines that are parallel to the same straight line are parallel to one another. (This is Proposition 30.)

2. Prove that the two versions of Pasch's axiom are equivalent.

3. Work through each of the following examples using Euclid's postulates and propositions (see Appendix A) to determine which steps are valid. Then find the "flaw" in each. [Examples a–c are reprinted from Dubnov (1963) with the permission of D.C. Heath and Company. Example d is reprinted from Maxwell (1961) with the permission of Cambridge University Press.]

Example a. A right angle is congruent to an obtuse angle.

Proof. From the endpoints of segment AB construct two congruent line segments AC and BD lying on the same side of AB so that $\angle DBA$ is a right angle and $\angle CAB$ is an obtuse angle. We shall prove that $\angle DBA \simeq \angle CAB$. Construct CD. Clearly, AB and CD are not parallel. Construct the perpendicular bisectors of segments AB and CD and let their point of intersection be N. Construct NA, NB, NC, and ND.

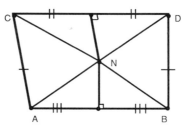

Figure 2.4

Case 1. The point N lies on the same side of AB as do C and D (Fig. 2.4). Clearly,

$$\triangle NAC \simeq \triangle NBD \quad \text{so} \quad \angle NAC \simeq \angle NBD \tag{1}$$

Furthermore,

$$\angle NAB \simeq \angle NBA \tag{2}$$

Thus by adding (1) and (2), $\angle CAB \simeq \angle DBA$.

Case 2. The point N lies on AB; that is, N is the midpoint of segment AB. Then, as in case 1, $\triangle NAC \simeq \triangle NBD$, and again $\angle NAC \simeq \angle NBD$. Thus by substitution $\angle BAC \simeq \angle ABD$.

Case 3. The point N lies on the opposite side of AB from C and D (Fig. 2.5). Then as in case 1, $\triangle NAC \simeq \triangle NBD$ and again

$$\angle NAC \simeq \angle NBD \tag{1}$$

Furthermore,

$$\angle NAB \simeq \angle NBA \tag{2}$$

Thus by subtracting (2) from (1), $\angle CAB \simeq \angle DBA$. $\square$

Example b. A rectangle inscribed in a square is also a square.

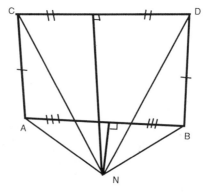

Figure 2.5

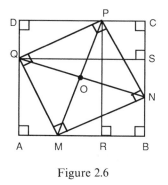

Figure 2.6

Proof. Let rectangle $MNPQ$ be inscribed in square $ABCD$ as shown in Fig. 2.6. Drop perpendiculars from P to AB and from Q to BC at R and S, respectively. Clearly $PR \simeq QS$. Furthermore $PM \simeq QN$. So $\triangle PMR \simeq \triangle QNS$, and hence $\angle PMR \simeq \angle QNS$. Consider quadrilateral $MBNO$ where O is the point of intersection of QN and PM. Its exterior angle at the vertex N is congruent to the interior angle at the vertex M, so that the two interior angles at the vertices N and M are supplementary. Thus the interior angles at the vertices B and O must also be supplementary. But $\angle ABC$ is a right angle and hence $\angle NOM$ must also be a right angle. Therefore the diagonals of rectangle $MNPQ$ are perpendicular. Hence $MNPQ$ is a square. $\qquad\square$

Example c. Two lines, exactly one of which is perpendicular to a third line, do not intersect.

Proof. At points A_0 and B_0 of the line l draw A_0Q and B_0P so that Q and P are on the same side of A_0B_0 and so that $\angle QA_0B_0$ is acute and $\angle PB_0A_0$ is a right angle (see Fig. 2.7). We shall prove that the rays A_0Q and B_0P do not intersect. Locate A_1 on A_0Q and B_1 on B_0P such that $d(A_1, A_0) = d(B_1, B_0) = \frac{1}{2}(d(A_0, B_0))$. Then for $i \geq 1$ locate points A_i on A_0Q such that A_i is between A_{i-1} and A_{i+1} and $d(A_{i+1}, A_i) = \frac{1}{2}(d(A_i, B_i))$. Also locate points B_i on B_0P such

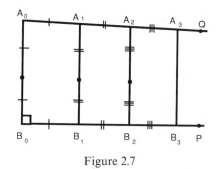

Figure 2.7

that B_i is between B_{i-1} and B_{i+1} and $d(B_{i+1}, B_i) = \frac{1}{2}(d(A_i, B_i))$. Clearly, for any i, segments $A_{i+1}A_i$ and $B_{i+1}B_i$ cannot have any points in common since if K were a common point, there would exist a triangle, $\triangle A_i K B_i$ in which the sum of two sides, $A_i K$ and $B_i K$, is less than or equal to the length of the third side, $A_i B_i$.   □

Example d. Every point inside a circle, other than the center, lies on its circumference.

Proof. Consider an arbitrary circle with center O and radius r and an arbitrary point $P \neq O$ inside it. Let Q be the point on OP such that P is between O and Q and such that $d(O, P) \cdot d(O, Q) = r^2$ (see Fig. 2.8). Let the perpendicular bisector of segment PQ at R intersect the circle at points U and V. Then

$$d(O, P) = d(O, R) - d(R, P)$$

and

$$d(O, Q) = d(O, R) + d(R, Q)$$
$$= d(O, R) + d(R, P).$$

So

$$\begin{aligned} d(O, P) \cdot d(O, Q) &= [d(O, R) - d(R, P)] \cdot [d(O, R) + d(R, P)] \\ &= d^2(O, R) - d^2(R, P) \\ &= [d^2(O, U) - d^2(R, U)] - [d^2(P, U) - d^2(R, U)] \\ &= d^2(O, U) - d^2(P, U) \\ &= d(O, P) \cdot d(O, Q) - d^2(P, U). \end{aligned}$$

Therefore $d(P, U) = 0$ so $P = U$.   □

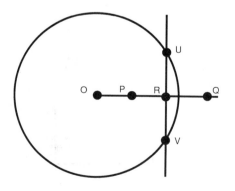

Figure 2.8

2.3. Non-Euclidean Geometry

The struggle to prove Euclid's fifth postulate, which began shortly after the appearance of *Elements* (ca. 300 B.C.) continued well into the 18th century, but eventually mathematicians realized that the fifth postulate is independent of the first four. In other words, there can exist geometries in which the negation of the fifth postulate is an axiom. These geometries came to be known as non-Euclidean. To initiate our study of these non-Euclidean geometries, we will consider the equivalent version of Euclid's fifth postulate, known as Playfair's axiom.

Postulate 5′ (Playfair's Axiom). Through a given point not on a given line can be drawn exactly one line not intersecting the given line.

Thus Euclid's geometry can be said to be based on Postulates 1–4 and 5′. Non-Euclidean geometry, on the other hand, is based on Euclid's Postulates 1–4 and a negation of Playfair's axiom. The two possible negations of Playfair's axiom given here lead to two vastly different non-Euclidean geometries— hyperbolic and elliptic (for elliptic geometry, a modification of Postulate 2 must also be made).

Hyperbolic Axiom. Through a given point, not on a given line, at least two lines can be drawn that do not intersect the given line.

Elliptic Axiom. Two lines always intersect.

The presentation of hyperbolic geometry that begins in the next section reflects the manner in which the subject developed historically. The proofs are based on Euclid's Postulates 1–4 and the hyperbolic axiom, with additional justification offered for the previously cited unstated assumptions of Euclid. This approach has the disadvantage that it lacks the rigorous nature of modern mathematics but it does expedite exploration of many of the unexpected results of this fascinating subject. Since the proofs of Euclid's Propositions 1–28 are based only on Postulates 1–4, we already have 28 theorems of hyperbolic geometry. With these theorems in hand, we are able to jump into the "heart" of plane hyperbolic geometry and develop almost immediately several interesting and "strange" results in this subject. The development of these results is, for the most part, surprisingly easy; although a few of the theorems necessary in the development are more difficult to prove than comparable theorems in Euclidean geometry. (More rigorous proofs for each of our results can be developed using Hilbert's axiom system with the hyperbolic axiom in place of Playfair's axiom.)

Many of the theorems we will encounter were developed by the Italian mathematician Gerolamo Saccheri (1667–1733) in his attempt to find a *reductio ad absurdum* proof of the fifth postulate. Influenced by the contemporary view that Euclidean geometry was the only possible geometry and faced with results radically different from those in Euclidean geometry, Saccheri allowed himself to think he had obtained a contradiction to the hyperbolic hypothesis after producing a long list of hyperbolic theorems. (He also rejected a second alternative hypothesis after much briefer consideration.) He recorded his work in a book with the intriguing title *Euclides ab Omni Naevo Vindicatus* (Euclid Freed of Every Flaw).

Credit for the discovery of hyperbolic geometry is generally given instead to the Russian mathematician Nicolai Ivanovich Lobachevsky (1793–1856) and the Hungarian mathematician János Bolyai (1802–1860), who published their independent work in 1829 and 1832, respectively. The eminent mathematician Karl Friedrich Gauss (1777–1855) also worked extensively in hyperbolic geometry but left his results unpublished. The details of the discoveries of these three men and the resistance they encountered provide one of the most fascinating episodes in the history of mathematics.

As the results of hyperbolic geometry unfold, the difficulty of visualizing these results within a world that most of us view as Euclidean becomes increasingly difficult. There are two frequently used geometric models that can aid our visualization of hyperbolic plane geometry. These are known as the Poincaré model and the Klein model (see Figs. 2.9 and 2.10). Both of these models assign interpretations to hyperbolic terms within the context of Euclidean geometry.

Poincaré Model

Hyperbolic Term	Interpretation
Point	Point interior to a given Euclidean circle $\mathscr{C}$
Line	Portion interior to $\mathscr{C}$ of diameter of $\mathscr{C}$ or a circle orthogonal to $\mathscr{C}$
Plane	Interior of $\mathscr{C}$

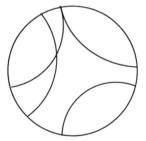

Figure 2.9. Poincaré model.

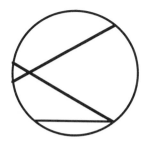

Figure 2.10. Klein model.

Under these interpretations the axioms and theorems of hyperbolic geometry become theorems of Euclidean geometry. The Poincaré model is important historically since it was used to demonstrate the consistency of hyperbolic geometry relative to Euclidean geometry.

Within this model, the measure of an angle determined as in Euclidean geometry is the measure of the hyperbolic angle. However, the correspondence between the hyperbolic and Euclidean distances is not nearly so straightforward. Essentially, Euclidean meter sticks must be viewed as getting longer as they are placed closer to the circumference of $\mathscr{C}$. A description of the life of "inhabitants" of the Poincaré model is given in Trudeau (1987, pp. 235–244).

The Klein model uses a similar interpretation of the term "point" but uses a more easily visualized interpretation of the term "line." However, in this model neither the distance nor the angle measures for hyperbolic geometry agree with their Euclidean counterparts.

Klein Model

Hyperbolic Term	Interpretation
Point	Point interior to a given Euclidean circle $\mathscr{C}$
Line	Open chord of $\mathscr{C}$
Plane	Interior of $\mathscr{C}$

This model will play an important role in Chapter 4, when we use Klein's definition of geometry to develop hyperbolic geometry as a subgeometry of the more general projective geometry.

The presentation of elliptic geometry, which ends this chapter uses an intuitive approach via spherical models rather than resorting to a drawn out axiomatic development. The aim of both presentations is to merely familiarize the reader with properties of non-Euclidean geometry.

2.4. Hyperbolic Geometry—Sensed Parallels

Since hyperbolic geometry results from the replacement of the fifth postulate by the hyperbolic axiom, we will begin our study of this geometry by determining the consequences of this new axiom. In doing so, we will immediately need to make use of one of the properties Euclid assumed without stating, namely, the continuity of lines. In the development that follows, we will accept Dedekind's axiom as an explicit statement of this property.

Dedekind's Axiom of Continuity. For every partition of the points on a line into two nonempty sets such that no point of either lies between two points of the other, there is a point of one set that lies between every other point of that set and every point of the other set.

Let P be a point and l a line not containing P as described in the hypothesis of the hyperbolic axiom. From P, construct a perpendicular to l at Q (12). Also construct a line m through P perpendicular to PQ at P (11). Let S be a second point on line m and construct QS (see Fig. 2.11). Then the points of QS can be partitioned into sets A and B as described herein.

(i) Let $A = \{X : X \text{ is on } QS \text{ and } PX \text{ intersects } l\}$.
(ii) Let $B = \{Y : Y \text{ is on } QS \text{ and } PY \text{ does not intersect } l\}$.

Clearly, Q is in set A and S is in set B (why?) so the sets are nonempty. And if X, X' are elements of A and Y, Y' are elements of B, Y cannot be between X and X' and X cannot be between Y and Y' (see Exercise 1). Thus by Dedekind's axiom there is some point T in either set A or set B such that T is between X and Y for all X in A and Y in B. It soon becomes apparent that T is in B (see Exercise 2). If a point R on QS moves along QS from Q to S then the line PR will rotate about the point P and $m \angle QPR$ will take on values from 0 to 90°. (In

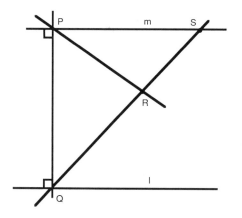

Figure 2.11

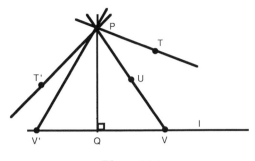

Figure 2.12

Fig. 2.11 this rotation would be counterclockwise.) Clearly, when R coincides with the point T given by Dedekind's axiom, $m \angle QPR = m \angle QPT < 90°$ and the line PT can be described as the first line in the rotation process that does not intersect l. Note that a similar situation arises on the other side of PQ; that is, there is another first line that does not intersect l, say PT'. For convenience we refer to these as the first lines on the *right* and *left* of PQ which do not intersect l. Furthermore, $\angle QPT \simeq \angle QPT'$. For if not, assume that $\angle QPT$ is greater than $\angle QPT'$ (see Fig. 2.12). Then construct PU so that $\angle QPU \simeq \angle QPT'$ where U is on the right side of PQ. Then since PT is the first line on the right of PQ, which does not intersect l, PU must intersect l at some point V. Let V' be a point on l to the left of PQ, such that segment $V'Q$ is congruent to segment VQ. Construct PV'. Since PQ is perpendicular to l, $\angle PQV' \simeq \angle PQV$. Thus $\angle QPV' \simeq \angle QPV$ (4) and therefore $\angle QPV' \simeq \angle QPT'$. So V' is on PT' and PT' intersects l. But this is a contradiction; therefore $\angle QPT \simeq \angle QPT'$.

Note that any of the infinite number of possible lines that lie between PT and PT' will not intersect l either.

The previous discussion is summarized in the following definition and theorem.

Definition 2.1. The first line through P relative to the counterclockwise (clockwise) rotation from PQ which does not intersect l is said to be *right- (left-) sensed parallel* to l through P. Any other line through P that does not intersect l is said to be *ultraparallel* to l through P or *nonintersecting* with l.

Theorem 29h. *If l is any line and P is any point not on l, then there are exactly two lines through P that do not intersect l and that make equal acute angles with the perpendicular from P to l, and that are such that every line through P lying within the angle containing that perpendicular intersects l, while every other line through P does not.*

Corollary. *Two lines with a common perpendicular are ultraparallel.*

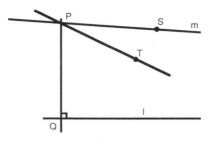

Figure 2.13

Before using this relation of sensed parallelism, some important properties of this relation must be established. Recall that in Euclidean geometry the parallelism relation satisfies the following properties:

1. If line l is the parallel to line m through P, then l is also the parallel to m through any other point R on l.
2. If line l is parallel to line m, then m is parallel to l (symmetry).
3. If line l is parallel to line m and m is parallel to line n, then l is parallel to n (transitivity).

The first of these merely states that the property of parallelism is independent of the point P. The second and third properties are the well-known properties of symmetry and transitivity as labeled. Before verifying that the relation of sensed parallelism in hyperbolic geometry satisfies these same properties, we shall first outline the procedure involved in proving that line l is sensed parallel to line m. From Definition 2.1 it should be apparent that in addition to verifying that l and m do not intersect, we must also show that at any point P on m, m is the first line in the rotation from the perpendicular to l that does not intersect l. In practice this second property is demonstrated by verifying in Fig. 2.13 that any line PT that intersects line m at P and enters $\angle QPS$ must intersect l. (Here PQ is perpendicular to l at Q and S is a point on m in the direction of parallelism from P.) This procedure is used in the proofs of the next three theorems. Each of these proofs is written for right-sensed parallels. The proofs for left-sensed parallels can be obtained by substituting the term "left" for "right" throughout.

Theorem 30h. *If a straight line l is the right- (left-) sensed parallel through a point P to a line m, it is at each of its points the right- (left-) sensed parallel to the line m.*

Proof. Assume l is right-sensed parallel to m through P. Let R be any other point on l.

Case 1. R is on the right side of P. Let PQ and RS be the perpendiculars to m at Q and $S(12)$. Let B be any point on l to the right of R. It is sufficient to show that every line RU lying within $\angle BRS$ must intersect m (Fig. 2.14). Construct

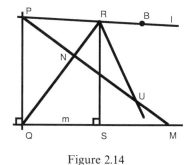

Figure 2.14

PU. Clearly, *PU* lies within ∠ *QPR* and hence must intersect *m* in a point *M* (definition of sensed parallels). Construct *QR*. By Pasch's axiom for △ *PQR*, *PU* must intersect segment *QR* in a point *N*. In △ *QNM*, *RU* intersects segment *NM*. But it does not intersect segment *QN*. Therefore, by Pasch's axiom, it must intersect segment *QM* and hence *m*.

Case 2. R is on the left side of *P*. (See Exercise 5.)   □

It can now be said that *l* is sensed parallel to *m* without specifying through which point, since this theorem shows the definition is independent of the point.

Theorem 31h. *If line l is right- (left-) sensed parallel to line m, then m is right- (left-) sensed parallel to l.*

Proof. Assume that *l* is right-sensed parallel to *m*. Let *P* be a point on *l*, *PQ* the perpendicular to *m* at *Q*, and *QR* the perpendicular to *l* at *R*.

Lemma. *R will be on the right side of PQ. (See Exercise 6.)*

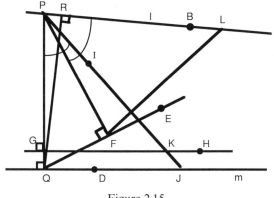

Figure 2.15

Let D be a point on m to the right of PQ. It is sufficient to show that every line QE lying within $\angle RQD$ must intersect l (see Fig. 2.15). Let PF be the perpendicular to QE at F. As before, the exterior angle theorem guarantees that F will lie to the right of PQ. Furthermore, segment PF is shorter than segment PQ (19). Thus on segment PQ there is a point G such that segment PG is congruent to segment PF (3). Draw GH perpendicular to PQ. Let B be a point on l such that B is to the right of R, and construct $\angle GPI \simeq \angle FPB$. Then PI will intersect m at J (definition of sensed parallel). Since GH intersects segment PQ in $\triangle PQJ$ and cannot intersect QJ it must intersect segment PJ at some point K. On line PB find a point L to the right of PQ such that segment PL is congruent to segment PK, and construct FL. Now $\triangle PGK \simeq \triangle PFL$ (4). So $\angle PFL$ is a right angle. But $\angle PFE$ is also a right angle. Hence $FE = FL$. Therefore QE intersects l at L.   □

Thus the relation of sensed parallelism is symmetric. The verification that this relation is also transitive is simplified by using the following lemma.

Lemma. *If m is right-sensed parallel to n, P and S are points on m (S to the right of P), and R is a point on n, then any line l entering $\angle RPS$ will intersect n at a point T on the right side of R.*

Proof. Let U be a point on line l below line m and let PQ be the perpendicular to n at Q.

Case 1. Q coincides with or lies to the left of R (Fig. 2.16). Then clearly $\angle QPU$ is less than $\angle QPS$ so PU intersects line n to the right of Q (definition of right-sensed parallel). Since PU cannot intersect segment QR (see Exercise 7), it must intersect n to the right of R.

Case 2. Q lies to the right of R (Fig. 2.17). Then line l will either enter $\triangle PQR$ and intersect side RQ (Pasch's axiom) and therefore line n as desired, or $\angle UPQ$ will be less than $\angle SPQ$ and thus l will intersect n to the right of R by the definition of right-sensed parallels.   □

Theorem 32h. *If two lines are both right- (left-) sensed parallel to a third line, then they are right- (left-) sensed parallel to one another.*

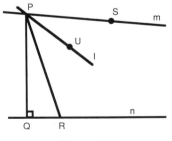

Figure 2.16

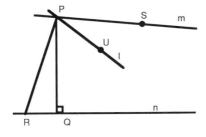

Figure 2.17

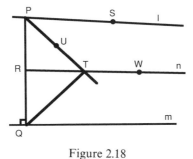

Figure 2.18

Proof. Assume that *l* is right-sensed parallel to *n* and *m* is right-sensed parallel to *n*. We shall show that *l* is right-sensed parallel to *m*.

Case 1. l and *m* lie on opposite sides of *n*. Clearly, *l* and *m* do not intersect since they lie on opposite sides of *n*. Let *P* and *S* be points on *l*, *S* to the right of *P*. Construct *PQ*, the perpendicular to *m* at *Q*. Then since *l* and *m* lie on opposite sides of *n*, *PQ* will intersect *n* at a point *R*. Let *PU* be a line entering ∠ *QPS*. It is sufficient to show that *PU* intersects *m* to the right of *Q* (see Fig. 2.18). Since *l* is right-sensed parallel to *n*, *PU* will intersect *n* at a point *T* by the preceding lemma. Construct line *TQ* and let *W* be a point on *n* to the right of *T*. Then line *PU* enters ∠ *QTW* at *T*. Since *m* is right-sensed parallel to *n*, *n* is right-sensed parallel to *m* by Theorem 31h, and the lemma can be used again to show that *PT* intersects *m* as desired.

Case 2. m is between *l* and *n*. Assume that *l* is not right-sensed parallel to *m* and let *P* be a point on *l* (see Fig. 2.19). By Theorem 29h there is a line *o* through *P* that is the right-sensed parallel to *m*. Since *m* is right-sensed parallel to *n*, *n* is right-sensed parallel to *m* (Theorem 31h). Furthermore *o* is right-sensed parallel to *m* and *o* and *n* lie on opposite sides of *m*. Therefore, by case 1, *o* is right-sensed parallel to *n*. This gives us two lines through *P*, both right-sensed parallel to *n*, contradicting the uniqueness guaranteed by Theorem 29h. Thus it follows that *l* is right-sensed parallel to *n*. □

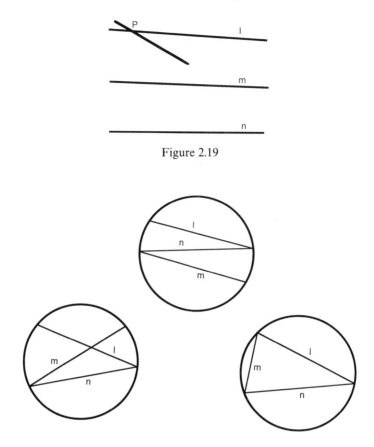

Figure 2.19

Figure 2.20

Hence the relation of sensed parallelism is transitive. Note, however, that the hypothesis of this theorem requires that the direction of parallelism be the same in both cases. But if, for example, l is right-sensed parallel to n and m is left-sensed parallel to n, l and m may *not* be sensed parallel. This is demonstrated using Klein models (see Fig. 2.20).

EXERCISES

1. Show that sets A and B described at the beginning of this section have the property that no point of either lies between two points of the other.

2. Verify that the point T guaranteed by Dedekind's axiom cannot be in set A and therefore must be in set B (where sets A and B are the sets described at the beginning of this section).

3. (a) Use a Klein model to show the right- and left-sensed parallels to a line l

through a point P not on l. (b) In the same model, show two lines through P that are ultraparallel to l.

4. (a) Use a Poincaré model to show the right- and left-sensed parallels to a line l through a point P not on l. (b) In the same model, show two lines through P that are ultraparallel to l.

5. Prove case 2 of Theorem 30h. [*Hint*: Choose U above l.]

6. Prove the lemma in the proof of Theorem 31h.

7. In case 1 of the proof of the lemma used to prove Theorem 32h, prove that PU cannot intersect segment QR. [*Hint*: You may need to refer to the separation of the plane by a line.]

2.5. Hyperbolic Geometry—Asymptotic Triangles

We will continue the study of sensed parallels by examining the figures formed by two sensed parallel lines and a transversal. Because these figures resemble triangles, they have come to be known as *asymptotic triangles* (see Fig. 2.21). In order to make use of the usual convention of naming a triangle by its three vertices, we will formalize the concept of ideal points. If l and m are sensed parallel lines, they are said to intersect in an *ideal* point. Ideal points will be represented by Greek letters, for example, Ω. Since there are right- and left-sensed parallels to every line, any line l will have exactly two ideal points. In the Poincaré and Klein models, the ideal points are the points lying on the circumference of $\mathscr{C}$. For convenience we will say that two sensed parallel lines l and m "intersect" at a point Ω, but we must be careful not to let this familiar terminology suggest that these ideal points possess the properties of ordinary hyperbolic points.

Definition 2.2. The figure consisting of two sensed parallel lines and a transversal intersecting the lines at A and B is referred to as an *asymptotic triangle*. If Ω is the ideal point determined by the sensed parallels, we refer to this asymptotic triangle as $\triangle AB\Omega$.

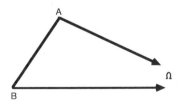

Figure 2.21. Asymptotic triangle, $\triangle AB\Omega$.

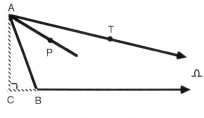

Figure 2.22

It is important to note that asymptotic triangles, despite the name, are *not* triangles so we cannot apply previous theorems about triangles to asymptotic triangles. However, asymptotic triangles do have some properties in common with triangles. In particular, Theorems 33h and 34h show that a modified Pasch's axiom holds for asymptotic triangles. Note that Theorem 33h is an extension of the lemma used in the proof of Theorem 32h, but here it is proved using the notation of asymptotic triangles.

Theorem 33h. *If a line passes within asymptotic triangle $\triangle AB\Omega$ through one of its vertices (including Ω) it will intersect the opposite side.*

Proof. Let AP be a line passing through A, and P, a point interior to $\triangle AB\Omega$. Let AC be the perpendicular to $B\Omega$ through A.

Case 1. AC coincides with AB or falls outside $\triangle AB\Omega$ (see Fig. 2.22). Then clearly $\angle PAC$ is smaller than $\angle TAC$ where T lies on side $A\Omega$. Hence AP intersects side $C\Omega$ (definition of sensed parallels). Since AP cannot intersect segment CB, it must intersect side $B\Omega$.

Case 2. AC lies within $\triangle AB\Omega$ (see Fig. 2.23). Then P may fall inside $\triangle ABC$ and hence AP intersects side BC (Pasch's axiom) and therefore side $B\Omega$, or P may fall inside $\triangle AC\Omega$ or on side AC. In this latter instance, as in case 1, AP must intersect side $C\Omega$ and hence side $B\Omega$.

The proof for a line through B is the same; so assume the line passes through Ω, that is, $P\Omega$ is sensed parallel to $A\Omega$ and $B\Omega$ (see Fig. 2.24). Construct AP. Then by the previous part of the proof AP intersects side $B\Omega$ at some point Q.

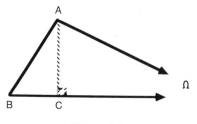

Figure 2.23

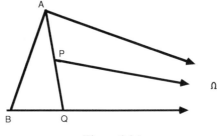

Figure 2.24

But $P\Omega$ intersects side AQ in $\triangle ABQ$ and therefore intersects side AB (Pasch's axiom).   □

Theorem 34h. *If a straight line intersects one of the sides of asymptotic triangle $\triangle AB\Omega$ but does not pass through a vertex (including Ω), it will intersect exactly one of the other two sides.*

Proof. See Exercise 2.   □

The analog of the exterior angle theorem for ordinary triangles (Proposition 16) can also be verified for asymptotic triangles. Note, however, that in an ordinary triangle each exterior angle has two opposite interior angles; whereas each exterior angle of an asymptotic triangle has only one.

Theorem 35h. *The exterior angles of asymptotic triangle $\triangle AB\Omega$ at A and B made by extending AB are greater than their respective opposite interior angles.*

Proof. Let AB be extended through B to C. It is sufficient to show that $\angle CB\Omega$ is greater than $\angle BA\Omega$. To do this, assume that the opposite is true, that is, $\angle CB\Omega$ is less than or equal to $\angle BA\Omega$. Through B construct BD such that D lies in the direction of parallelism from AB and $\angle CBD \simeq \angle BA\Omega$.

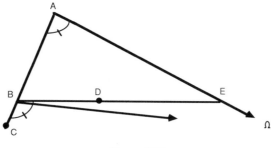

Figure 2.25

Case 1. *D* lies inside $\triangle AB\Omega$ (see Fig. 2.25). Then by Theorem 33h, *BD* intersects $A\Omega$ at some point *E*. But then in $\triangle ABE$ the exterior angle at *B* is congruent to the interior angle at *A*, which contradicts Proposition 16.

Case 2. *D* lies on $B\Omega$ (see Fig. 2.26). Let *M* be the midpoint of segment *AB* (10). Construct *MN* perpendicular to $B\Omega$ at *N*. Clearly, *N* cannot coincide with *B* (why?). We shall assume that *N* falls to the right of *B* (if $A\Omega$ and $B\Omega$ are right-sensed parallel as shown in Fig. 2.26). The proof for the case when *N* falls to the left of *B* is similar (see Exercise 3). Extend $A\Omega$ to *L* so the segment *LA* is congruent to segment *BN*. Construct *ML*. Then $\angle LAM \simeq \angle NBM$ since they are supplements of congruent angles. Hence $\triangle LAM \simeq \triangle NBM$, and $\angle BMN \simeq \angle AML$. Therefore $LM = MN$. Furthermore, $\angle ALM \simeq \angle BNM$. So $\angle ALM$ is a right angle. Thus $A\Omega$ is ultraparallel to $B\Omega$. But this contradicts the hypothesis. Thus both cases lead to a contradiction, and hence it follows that $\angle CB\Omega$ is greater than $\angle BA\Omega$.   □

Note that case 2 of this proof demonstrates the following theorem.

Theorem 36h. *Two lines cut by a transversal so as to make alternate angles congruent are ultraparallel.*

As a result of this theorem, Euclid's Propositions 27 and 28 refer to ultraparallel lines.

The familiar triangle congruence theorems of Euclidean geometry also have their analogs in hyperbolic geometry. Here, since two of the three sides of an asymptotic triangle are infinite, there are only two angles and one side to consider. In other words, two asymptotic triangles are said to be congruent whenever their finite sides and the two pairs of corresponding angles are congruent.

Theorem 37h. *If segment AB is congruent to segment A'B' and $\angle BA\Omega$ is congruent to $\angle B'A'\Omega'$ in asymptotic triangles, $\triangle AB\Omega$ and $\triangle A'B'\Omega'$, then $\angle AB\Omega$ is congruent to $\angle A'B'\Omega'$.*

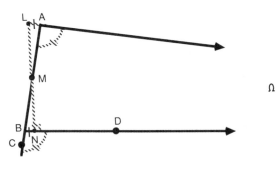

Figure 2.26

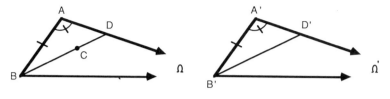

Figure 2.27

Proof. Assume $\angle ABΩ \not\simeq \angle A'B'Ω'$; in particular assume that $\angle ABΩ$ is greater than $\angle A'B'Ω'$. Let C be a point in the direction of parallelism from AB such that $\angle ABC \simeq \angle A'B'Ω'$ (see Fig. 2.27). By Theorem 33h, BC intersects side $AΩ$ at some point D. On $A'Ω'$ find D' such that segment AD is congruent to segment $A'D'$. Construct $B'D'$. Then $\triangle ABD \simeq \triangle A'B'D'$, so $\angle A'B'D' \simeq \angle ABD$. But $\angle ABD \simeq \angle A'B'Ω'$ and thus $\angle A'B'D' \simeq \angle A'B'Ω'$. And $B'D' = B'Ω'$. But this is a contradiction so $\angle ABΩ \simeq \angle A'B'Ω'$. ☐

Two other congruence theorems for asymptotic triangles are stated here (see Exercises 4 and 5).

Theorem 38h. *In asymptotic triangles* $\triangle ABΩ$ *and* $\triangle A'B'Ω'$, *if* $\angle BAΩ \simeq \angle B'A'Ω'$ *and* $\angle ABΩ \simeq \angle A'B'Ω'$, *then segment AB is congruent to segment $A'B'$.*

Theorem 39h. *In asymptotic triangles* $\triangle ABΩ$ *and* $\triangle A'B'Ω'$, *if segment AB is congruent to segment $A'B'$,* $\angle ABΩ \simeq \angle BAΩ$, *and* $\angle A'B'Ω' \simeq \angle B'A'Ω'$, *then* $\angle ABΩ \simeq \angle A'B'Ω' \simeq \angle BAΩ \simeq \angle B'A'Ω'$.

These asymptotic triangle theorems lead to a unique concept in hyperbolic geometry, namely, the *angle of parallelism*. The definition of this concept uses a mapping on the set of positive real numbers.

Let PQ be a segment of length h. Let QS be the line perpendicular to PQ at Q and PR the line sensed parallel to QS through P (see Fig. 2.28). Then $a(h) = m\angle QPR$ where $m\angle QPR$ denotes the measure of $\angle QPR$.

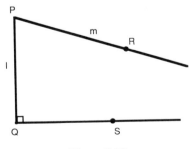

Figure 2.28

Theorems 37h and 29h can be used to show that the mapping $a(h)$ is well defined. As shown in Exercise 6, this mapping is also one-to-one and order reversing [i.e., if $h < h'$, then $a(h) > a(h')$]. Furthermore, it can be shown that $a(h)$ is a continuous mapping. These results are summarized in the following theorem.

Theorem 40h. *The mapping $a(h)$ just described is continuous, one to one, and order reversing.*

Definition 2.3. An angle with measure $a(h)$ is called an *angle of parallelism of h.*

In *A Survey of Geometry* (pp. 414–416), Howard Eves (1972) demonstrates that

$$a(h) = 2 \arctan (e^{-h})$$

if the unit of length is chosen as the distance corresponding to the angle of parallelism $a = 2 \arctan (e^{-1})$.

This leads to another interesting property of hyperbolic geometry that is not possessed by Euclidean geometry. Note that in both Euclidean and hyperbolic geometry, angles possess a natural unit of measure that can be geometrically constructed since right angles can be constructed. Because of this, angles are said to be *absolute* in both geometries. In Euclidean geometry, lengths are not absolute; since there is no natural unit of length structurally connected with the geometry. However, in hyperbolic geometry lengths are absolute because the mapping $a(h)$ associates to any angle (e.g., 45°) a definite distance h; and once an angle of measure 45° is constructed the corresponding angle of parallelism can be constructed. [Note that this statement assumes that it is possible to *construct* a line perpendicular to one of two intersecting lines and sensed parallel to the other; that is, if lines l and m intersect at point P as shown in Fig. 2.28, it is possible to construct the line QS perpendicular to l at Q and sensed parallel to m. This construction is demonstrated in Wolfe (1945, pp. 97–99).]

EXERCISES

1. Explain why there are no more than two ideal points on a hyperbolic line.

2. Prove Theorem 34h.

3. Complete the verification of case 2 in the proof of Theorem 35h by considering the case where N falls to the left of B.

4. Prove Theorem 38h.

5. Prove Theorem 39h.

6. Prove: If $h < h'$, then $a(h) > a(h')$.

7. Prove that the sum of the measures of the two angles at the ordinary vertices of an asymptotic triangle is less than 180°. [*Hint:* Use Theorem 35h.]

2.6. Hyperbolic Geometry—Saccheri Quadrilaterals

A second figure of importance in hyperbolic geometry is the Saccheri quadrilateral (see Fig. 2.29) in honor of the efforts of Gerolamo Saccheri who almost discovered non-Euclidean geometry.

Definition 2.4. A *Saccheri quadrilateral* is a quadrilateral *ABCD* with two adjacent right angles at *A* and *B* and with sides $AD \simeq BC$. Side *AB* is called the *base* and side *DC* is called the *summit*.

We shall soon see that one of the implications of the hyperbolic axiom is that the angles at *C* and *D* in this figure are not right angles as they are in Euclidean geometry. There are, however, properties of Saccheri quadrilaterals common to both Euclidean and hyperbolic geometry since their proofs require only results based on Euclid's first four postulates. Two of these common properties are stated in Theorem 41h and conclusion (1) of Theorem 42h.

Theorem 41h. *The line joining the midpoints of the base and summit of a Saccheri quadrilateral is perpendicular to both of them.*

Proof. See Exercise 1.

Corollary. *The base and summit of a Saccheri quadrilateral are ultraparallel.*

Theorem 42h. *The summit angles of a Saccheri quadrilateral are* (1) *congruent and* (2) *acute.*

Proof. See Exercise 3.

As indicated earlier, the proof of part (2) depends on the hyperbolic axiom. In Euclidean geometry this conclusion must be changed to "right," whereas in elliptic geometry this conclusion must be changed to "obtuse." In fact, Theorem 42h is equivalent to the hyperbolic axiom whereas the Euclidean version is equivalent to Euclid's parallel postulate. This theorem also leads to one of the dramatic results in hyperbolic geometry, namely, that the angle sum

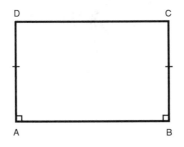

Figure 2.29

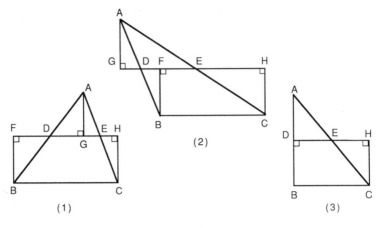

Figure 2.30

of every triangle is less than 180°. As we shall see in the next section, the angle sum is *not* even constant for all triangles.

Theorem 43h. *The sum of the angles of every triangle is less than two right angles.*

Proof. Assume △*ABC* is an arbitrary triangle with base *BC*. Let *D* and *E* be the midpoints of sides *AB* and *AC*, respectively. And let *BF*, *AG*, and *CH* be the perpendiculars to *DE* from *B*, *A*, and *C*. Then, as shown in Exercise 5, there are three possible cases (see Fig. 2.30).

Case 1. Since ∠ *BDF* ≃ ∠ *ADG* (15), it follows that △*BDF* ≃ △*ADG* and thus ∠ *FBD* ≃ ∠ *GAD* and segment *BF* is congruent to segment *AG* (26). Likewise, ∠ *HCE* ≃ ∠ *GAE* and segment *AG* is congruent to segment *CH*. Hence segment *BF* is congruent to segment *CH*, and quadrilateral *BFHC* is a Saccheri quadrilateral. Thus by Theorem 42h, ∠ *FBC* ≃ ∠ *HCB* and both are acute, so their sum is less than two right angles. But

$$\angle FBC + \angle HCB = \angle FBD + \angle DBC + \angle HCE + \angle ECB$$
$$= \angle GAB + \angle ABC + \angle GAE + \angle ACB$$
$$= \angle ABC + \angle BAC + \angle ACB.$$

It follows that the angle sum of △ *ABC* is less than two right angles.
Cases 2 *and* 3. See Exercise 6.   □

In the preceding proof, △*ABC* is said to be *equivalent* to Saccheri quadrilateral *BFHC*.

Several important results are immediate corollaries of this theorem.

Corollary 1. *The sum of the angles of a quadrilateral is less than four right angles.*

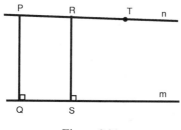

Figure 2.31

Corollary 2. *Two lines cannot have more than one common perpendicular.*

Corollary 3. *There do not exist lines that are everywhere equidistant.*

As Corollary 3 states, lines are never equidistant. Instead the distance between sensed parallels varies from point to point as shown in the following theorem.

Theorem 44h. *The perpendicular distance from a point on one of two sensed parallels to the other line decreases as the point moves in the direction of parallelism.*

Proof. Let lines n and m be right-sensed parallel. Choose points P and R on n (see Fig. 2.31). Construct PQ and RS perpendicular to m from P and R, respectively. (Assume R is to the right of PQ.) Then it suffices to show that $m(RS) < m(PQ)$. Let T be to the right of R. Now $m(\angle PRS) + m(\angle SRT) = 180°$, and $m(\angle QPR) + m(\angle PRS) < 180°$ by Corollary 1 of Theorem 43h. Thus $m(\angle QPR) < m(\angle SRT)$ so $m(PQ) = a^{-1}(m\angle QPR) > a^{-1}(m\angle SRT) = m(RS)$. □

Theorem 43h yields still another result that is vastly different from what happens in Euclidean geometry.

Theorem 45h. *If the three angles of one triangle are congruent respectively to three angles of a second triangle, then the triangles are congruent.*

Proof. Let $\triangle ABC$ and $\triangle A'B'C'$ be two triangles with corresponding angles congruent. Now if any pair of corresponding sides is congruent, then the triangles are congruent (Proposition 26). Hence, assume that none of the three pairs of corresponding sides is congruent. In particular, assume that $AB \not\simeq A'B'$; assume furthermore that $m(AB) > m(A'B')$. So find A'' on AB such that $A''B \simeq A'B'$; and on BC find C'' such that $BC'' \simeq B'C'$ (see Fig. 2.32). Now $\triangle A''BC'' \simeq \triangle A'B'C'$ by Proposition 4. So $\angle BA''C'' \simeq \angle B'A'C'$ and $\angle BC''A'' \simeq \angle B'C'A'$. Thus $\angle BA''C'' \simeq \angle BAC$ and $\angle BC''A'' \simeq \angle BCA$.

Case 1. C is between B and C''. Then by Pasch's axiom, $A''C''$ intersects AC

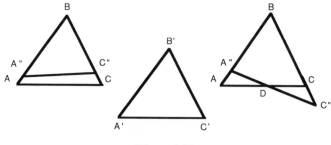

Figure 2.32

at a point D. And in $\triangle DCC''$, $\angle CC''D \simeq \angle BCD$. But $\angle BCD$ is an exterior angle and this is a contradiction to the exterior angle theorem.

Case 2. C'' is between B and C. Then $A''ACC''$ is a quadrilateral and

$$m \angle C''A''A + m \angle A''AC + m \angle ACC'' + m \angle CC''A''$$
$$= (180 - m \angle BA''C'') + m \angle A''AC + m \angle ACC'' + (180 - m \angle BC''A'')$$
$$= 180 - m \angle A''AC + m \angle A''AC + m \angle ACC'' + 180 - m \angle C''CA$$
$$= 360.$$

But this contradicts Corollary 1 of Theorem 43h. Therefore $\triangle ABC \simeq \triangle A'B'C'$.

Recall that in Euclidean geometry, two triangles are said to be *similar* if there is a one-to-one correspondence between the vertices of the two triangles such that corresponding angles are congruent and the lengths of corresponding sides are proportional. However, the preceding theorem indicates that in hyperbolic geometry any two triangles satisfying these properties are automatically congruent. Thus we do not have any similar, but noncongruent triangles in hyperbolic geometry.

EXERCISES

1. Prove Theorem 41h.

2. Prove the corollary of Theorem 41h.

3. Prove Theorem 42h. [Hint: To prove (2), construct right-sensed parallels to AB at C and D and apply Theorem 35h to asymptotic triangle $\triangle CD\Omega$.]

4. Prove that Theorem 42h is equivalent to the hyperbolic axiom.

5. Show that in $\triangle ABC$, where D and E are the midpoints of AB and AC, respectively, the perpendiculars to line DE from A and B must either coincide with or lie on opposite sides of AB. (Thus there are only the three possible cases as shown in Fig. 2.30).

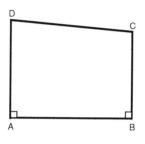

Figure 2.33

6. Prove cases 2 and 3 of Theorem 43h.

7. Why can there be no squares or rectangles in hyperbolic geometry?

8. Show that in Fig. 2.33, if $AD > BC$, then $m(\angle BCD) > m(\angle ADC)$.

2.7. Hyperbolic Geometry—Area of Triangles

In the previous section we discovered that in hyperbolic geometry the angle sum of every triangle is less than 180° and that similar triangles do not exist. We now show that in this geometry the area of a triangle is determined by its angle sum. However, before proceeding with the necessary theorems it is prudent to recall the axioms that any area function must satisfy.

Area Axioms

A1. The area of any set must be nonnegative.
A2. The area of congruent sets must be the same.
A3. The area of the union of disjoint sets must equal the sum of the areas of the sets.

To begin the sequence of theorems that will lead to the desired result, we need to return to a consideration of Saccheri quadrilaterals.

Theorem 46h. *Two Saccheri quadrilaterals with congruent summits and summit angles are congruent.*

Proof. Let $ABCD$ and $EFGH$ be two Saccheri quadrilaterals with $AB \simeq EF$ and $\angle DAB \simeq \angle HEF \simeq \angle EFG \simeq \angle ABC$. We must show that $AD \simeq EH$ (consequently $BC \simeq FG$) and that $DC \simeq HG$.

Part 1. $AD \simeq EH$. Assume that this is not true; in particular, assume that $m(AD) < m(EH)$. Find H' on EH and G' on FG so that $EH' \simeq AD$ and $FG' \simeq$

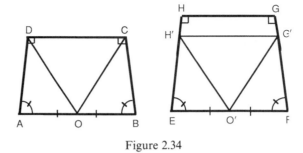

Figure 2.34

BC. Construct $H'G'$ (see Fig. 2.34). Let O and O' be the midpoints of AB and EF, respectively. Construct DO, CO, $H'O'$, and $G'O'$. Clearly, $\triangle DAO \simeq \triangle H'EO'$, and $\triangle OCB \simeq \triangle O'G'F$. Thus $DO \simeq H'O'$ and $CO \simeq O'G'$ and $\angle DOC \simeq \angle H'O'G'$. So $\triangle DOC \simeq \triangle H'O'G'$. Then $\angle EH'G' \simeq \angle ADC$ and both are right angles. Likewise $\angle FG'H' \simeq \angle BCD$ and both are right angles. Therefore $\angle HH'G'$ and $\angle GG'H'$ are also right angles. Thus quadrilateral $HH'G'G$ has four right angles, contradicting Corollary 1 of Theorem 43h. So $AD \simeq EH$.

Part 2. $DC \simeq HG$. see Exercise 3.   □

With this result, we can prove a specialized version of the general area theorem for triangles.

Theorem 47h. *Two triangles with the same angle sum and one pair of congruent sides have the same area.*

Proof. Let $\triangle ABC$ and $\triangle DEF$ be two triangles with the same angle sum and assume $AB \simeq DE$. Let G and H be the midpoints of AC and BC. Construct GH. Let I, J, K be the feet of the perpendiculars to GH from A, C, and B, respectively. As in the proof of Theorem 43h, there are three possible cases. In the case shown in Fig. 2.35, $\triangle AIG \simeq \triangle CJG$ and $\triangle CJH \simeq \triangle BKH$. So $IA \simeq KB$ and $AIKB$ is a Saccheri quadrilateral. Then clearly area($AIKB$) = area($\triangle ABC$) by the area axioms.

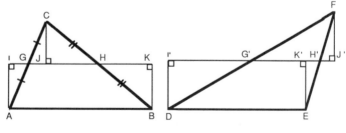

Figure 2.35

Furthermore $m(\angle IAB) = m(\angle CAB) + m(\angle GCJ)$ and $m(\angle KBA) =$ $m(\angle CBA) + m(\angle HCJ)$. Also $m(\angle CAB) + m(\angle CBA) + m(\angle GCJ) +$ $m(\angle HCJ) = m(\angle CAB) + m(\angle CBA) + m(\angle ACB)$. Therefore, since $AIKB$ is a Saccheri quadrilateral, $m(\angle IAB) = m(\angle KBA) = \frac{1}{2}[m(\angle CAB) + m(\angle CBA) + m(\angle ACB)]$.

As shown by Exercise 4, similar proofs for the other two cases demonstrate that a triangle and its equivalent Saccheri quadrilateral always have the same area and furthermore the angle sum of the triangle equals the sum of the summit angles of the equivalent Saccheri quadrilateral.

Completing the same construction on DEF creates the Saccheri quadrilateral $I'DEK'$ with area$(I'DEK')$ = area$(\triangle DEF)$ and with $m(\angle I'DE) =$ $m(\angle DEK') = \frac{1}{2}(m(\angle FDE) + m(\angle FED) + m(\angle DFE))$. But since the summit angles of the two Saccheri quadrilaterals are congruent from the preceding and the hypothesis, and since $AB \simeq DE$, it follows by Theorem 46h that $I'DEK \simeq AIKB$. So area$(I'DEK')$ = area$(AIKB)$ and hence area$(\triangle ABC) =$ area$(\triangle DEF)$.   □

In order to prove Theorem 48h, the generalized version of the previous theorem, we first demonstrate the following result.

Lemma. *In $\triangle ABC$ if FE is perpendicular to the perpendicular bisector of BC and intersects AC at its midpoint, it will also intersect AB at its midpoint.*

Proof. As in the proof of Theorem 43h there again are three cases. For the first case Fig. 2.36 can be used to complete the proof and similar arguments can be used in the other two cases.   □

Theorem 48h. *Any two triangles with the same angle sum have the same area.*

Proof. Let $\triangle ABC$ and $\triangle A'B'C'$ be two triangles with the same angle sum. Without loss of generality, assume that $m(A'C') > m(AC)$. (Note that if any pair of sides is congruent, the result follows immediately from Theorem 47h.) As in

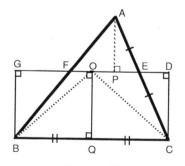

Figure 2.36

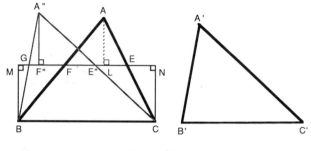

Figure 2.37

the proof of Theorem 47h, construct the Saccheri quadrilateral on BC (see Fig. 2.37). Then let E'' be on FE so that $m(CE'') = \frac{1}{2}m(A'C')$. E'' will not coincide with E or N since $\frac{1}{2}m(A'C') > \frac{1}{2}m(AC) > m(CN)$. Construct CE'' and extend it to a point A'' so that $E''A'' \simeq CE''$. Construct $A''B$. Now let F'' be the foot of the perpendicular to MN from A''. Since FE is perpendicular to the perpendicular bisector of BC by Theorem 41h and intersects $A''C$ at its midpoint, it will also intersect $A''B$ at its midpoint G by the previous lemma. Therefore $\triangle BMG \simeq \triangle A''F''G$ and $\triangle A''E''F'' \simeq \triangle CE''N$. So area$(\triangle A''BC) =$ area$(MBCN)$. But as in the proof of Theorem 47h, area$(\triangle ABC) =$ area$(MBCN)$. And so area$(\triangle A''BC) =$ area$(\triangle ABC)$. Furthermore, as in the proof of Theorem 47h, the angle sum of $\triangle A''BC = m(\angle MBC) + m(\angle BCN) =$ angle sum of $\triangle ABC$. Therefore the angle sum of $\triangle A''BC =$ angle sum of $\triangle A'B'C'$, and $A''C \simeq A'C'$. So by Theorem 47h, area$(\triangle A''BC) =$ area$(\triangle A'B'C')$ and thus area$(\triangle ABC) =$ area$(\triangle A'B'C')$.

Thus unlike Euclidean geometry, where the area of a triangle is determined by the lengths of its base and altitude, the preceding theorem demonstrates that the area of a triangle in hyperbolic geometry is completely determined by its angle sum. The relation between the area and the angle sum for triangles is stated in terms of the defect of a triangle.

Definition 2.4. The (*angular*) *defect* of a triangle is the numerical difference, 180 − the angle sum of the triangle; that is, the angular defect of $\triangle ABC = 180 - [m(\angle ABC) + m(\angle BCA) + m(\angle CAB)]$.

Theorem 49h. *If a triangle is divided into two triangles by a line from a vertex to a point on the opposite side, the defect of the original triangle is equal to the sum of the defects of the two smaller triangles.*

Proof. See Exercise 6.   □

Theorem 48h implies that the area of a triangle can be considered either as a function of the angle sum of the triangle or as a function of the angular defect of the triangle. From Theorem 49h and the axioms of area, it follows that the area function A must preserve addition of angular defects. Since this function A must be a continuous function, then a result of elementary calculus says that there is a constant k such that

$$A(\triangle ABC) = k^2(\text{defect}(\triangle ABC)).$$

This result is summarized in the following theorem. One proof of this theorem is credited to Gauss (Coxeter, 1969); another proof can be found in Moise (1974).

Theorem 50h. *There is a constant k such that the* area$(\triangle ABC) = k^2\{180 - [m(\angle ABC) + m(\angle BCA) + m(\angle CAB)]\}$.

EXERCISES

1. Using only Postulates 1 through 4 (and Propositions 1 through 28) prove the following. If the sum of the angles of a triangle is the same for all triangles then that sum is 180°. [*Hint*: Consider a triangle partitioned by a line joining a vertex with a point on the opposite side.] What does this result say about triangles in hyperbolic geometry?

2. The following "proof" of the existence of a triangle with angle sum equal to 180° is reprinted from Dubnov's (1963) *Mistakes in Geometric Proofs* with the permission of D.C. Heath and Co. (a) Does this "proof" make use of the parallel postulate? (b) What is wrong with the proof?

Claim. There exists a triangle with angle sum equal to 180°.

Proof. Since the angle sum of a triangle is less than or equal to 180°, let $\triangle ABC$ (see Fig. 2.38) be a triangle with the greatest angle sum, call this sum a. We shall prove that $a = 180°$.
   $\angle 1 + \angle 2 + \angle 6 \le a$ and $\angle 3 + \angle 4 + \angle 5 \le a$ (Why?) So $\angle 1 + \angle 2 + \angle 3 + \angle 4 + \angle 5 + \angle 6 \le 2a$. But $\angle 5 + \angle 6 = 180°$, and $\angle 1 + \angle 2 + \angle 3 + \angle 4 = a$. So $a + 180° \le 2a$ or $a \ge 180°$. Thus $a = 180°$.

3. Verify part 2 in the proof of Theorem 46h.

4. Prove that a triangle and its equivalent Saccheri quadrilateral have the

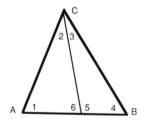

Figure 2.38

same area and that the angle sum of the triangle equals the sum of the
summit angles of the equivalent Saccheri quadrilateral.

5. Prove the lemma used in the proof of Theorem 48h.

6. Prove Theorem 49h.

7. Prove: The angle sum of a convex polygon of n sides is less than $(n-2)180°$.

2.8. Hyperbolic Geometry—Ultraparallels

In this final section on hyperbolic geometry, we will briefly consider the second
type of parallel lines, namely, ultraparallels. Recall that if l is a line and P a
point not on l then a line m through P is said to be ultraparallel to l if l and m do
not intersect and are not sensed parallel. As in the case of sensed parallels, the
definition of ultraparallelism is independent of the point P and the relation is
symmetric. These properties are formalized in the following theorems, which
can be verified by indirect proofs (see Exercises 2 and 3).

Theorem 51h. *If a line is ultraparallel through a given point to a given line, it is at
each of its points, ultraparallel to the given line.*

Theorem 52h. *If one line is ultraparallel to a second, then the second is
ultraparallel to the first.*

However, unlike sensed parallelism, ultraparallelism is *not* transitive. In
terms of line l and point P, any line lying within the vertical angles formed by
the sensed parallels to l through P is ultraparallel to l. In particular, any two of
these lines, say m and n, are both ultraparallel to l but m and n are not
ultraparallel since they intersect at P (see Fig. 2.39).

Two familiar properties in Euclidean geometry are (1) two parallel lines
have an infinite number of common perpendiculars; and (2) the (per-

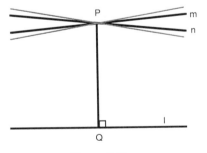

Figure 2.39

pendicular) distance between two parallel lines is constant (i.e., parallel lines are everywhere equidistant). In hyperbolic geometry, we have already observed that sensed parallel lines do not have common perpendiculars and that the perpendicular distance between sensed parallel lines decreases in the direction of parallelism (Theorem 44h). Furthermore one of the corollaries of Theorem 43h demonstrates that two ultraparallel lines do *not* have more than one common perpendicular. That a common perpendicular between two ultraparallel lines does exist is verified by the following theorem. Since the proof of this theorem is somewhat involved you may find it enlightening to sketch the specific points and lines one by one as you encounter them in your reading.

Theorem 53h. *Two ultraparallel lines have a common perpendicular.*

Proof. Let n and m be ultraparallel. Let A and B be any two points on n and construct AC and BD perpendicular to m at C and D. Now if segments AC and BD are congruent, $ABCD$ is a Saccheri quadrilateral (see Fig. 2.40) and the common perpendicular is the line connecting the midpoints of AB and CD. If AC and BD are not congruent, assume that $m(AC) > m(BD)$. Find E on AC such that $CE \simeq BD$. At E draw EF on the side of AC determined by BD such that $\angle CEF \simeq \angle DBG$ where G is a point on n such that B is between A and G. If will be shown that EF intersects n. Let $C\Omega$ and $D\Omega$ be sensed parallel to n in the direction from A to B. Let H be a point on m such that D is between C and H. Then $C\Omega$ contains points in the interior of $\angle ACH$, and $D\Omega$ likewise contains points in the interior of $\angle BDH$ since m is ultraparallel to n. Now $m(\angle HD\Omega) > m(\angle HC\Omega)$ by the exterior angle theorem for asymptotic triangles. Construct CJ such that $\angle JCH \simeq \angle \Omega DH$. Then CJ will intersect n at a point O. Now since $EC \simeq BD$, $\angle FEC \simeq \angle GBD$, and $\angle ECJ \simeq \angle BD\Omega$, EF is sensed parallel to CJ and hence cannot intersect segment CO. Therefore EF must intersect segment AO in a point K. Construct KL perpendicular to m. On n and m on the side of BD opposite A, find M and N such that $BM \simeq EK$, and $DN \simeq CL$. Construct MN. Then $\triangle ECL \simeq \triangle BDN$ and consequently $\triangle EKL \simeq \triangle BMN$ and thus $KL \simeq MN$. Furthermore, $m(\angle DNM) =$

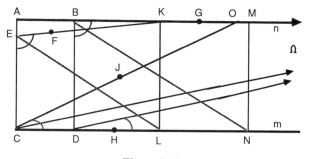

Figure 2.40

$m(\angle DNB) + m(\angle BNM) = m(\angle CLE) + m(\angle ELK) = 90$. So $KMNL$ is a Saccheri quadrilateral and the line joining the midpoints of segments KM and LN is perpendicular to both n and m.   □

Corollary. *Two ultraparallel lines have exactly one common perpendicular.*

Note that the proof of the previous theorem demonstrates that the common perpendicular exists but the actual construction cannot be accomplished without the ability to construct sensed parallels. This construction can be done and is demonstrated in Wolfe (1945).

The question of the distance between ultraparallel lines can now be answered in terms of the unique common perpendicular. The proof of this theorem is left for the exercises.

Theorem 54h. *If X is an arbitrary point on m, a line ultraparallel to n, then the perpendicular distance from X to n is a minimum along the common perpendicular to m and n (see Fig. 2.41).*

With this theorem, we conclude our introduction to hyperbolic geometry. The approach we have taken is similar to the historic development of the subject; that is, we started with Euclid's five postulates and replaced the fifth postulate (in the form of Playfair's axiom) by a negation in the form of the hyperbolic axiom. With this one change, we have obtained a new geometry with several "strange" properties that make it radically different from Euclidean geometry. In the next section, we will explore the geometry that results when we replace Euclid's fifth postulate by a second possible negation.

EXERCISES

1. Sketch each of the following in a Klein model. (Draw one model for each.)
   (a) Two intersecting lines that are both sensed parallel to a third line.
   (b) Two intersecting lines that are both ultraparallel to a third line. (c) Two sensed parallel lines that both intersect a third line. (d) Two sensed parallel lines that are both sensed parallel to a third line. (e) Two sensed parallel lines that are both ultraparallel to a third line. (f) Two ultraparallel lines

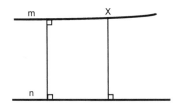

Figure 2.41

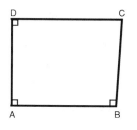

Figure 2.42

that both intersect a third line. (g) Two ultraparallel lines that are both ultraparallel to a third line.

2. Prove Theorem 51h.

3. Prove Theorem 52h.

4. Without using Theorem 54h, prove that in Fig. 2.42, $DC > AB$. (Any such quadrilateral with three right angles is known as a *Lambert quadrilateral*.)

5. Using the result of Exercise 4, prove that the summit of a Saccheri quadrilateral is greater than the base.

6. Using the result of Exercise 4, prove Theorem 54h.

2.9. Elliptic Geometry

The consequences of the hyperbolic axiom had been thoroughly explored before the systematic study of elliptic geometry began. The initiation of the study of this second non-Euclidean geometry can be traced to 1854 when G.B.F. Riemann gave an inaugural lecture at the University of Gottingen entitled "On the Hypotheses Which Underlie the Foundations of Geometry."

As with hyperbolic geometry, an axiomatic system for elliptic geometry is obtained from Euclid's geometry by replacing the fifth postulate in the form of Playfair's axiom with a negation. In this case the negation is known as the elliptic axiom.

Elliptic Axiom. Two lines always intersect.

Unfortunately, it soon becomes evident that the axiomatic system consisting of this axiom and Euclid's first four postulates is not consistent, since the first four postulates imply the validity of Proposition 27, which asserts the existence of parallel (i.e., nonintersecting) lines. In order to obtain a consistent system that contains the elliptic axiom and maintains as many of the properties of Euclidean geometry as possible, Euclid's proof of Proposition 27 must be invalidated. An examination of this proof (see Section 2.2) shows that it makes use of Proposition 16, but in the proof of this latter proposition, Euclid inferred from Postulate 2 the infinite extent of a line. If Postulate 2 is interpreted as saying only that a line is boundless but *not* necessarily of infinite extent, the proof of Proposition 16 and therefore the proof of Proposition 27 becomes invalid.

Thus to obtain a consistent non-Euclidean geometry containing the elliptic axiom, Euclid's second postulate must be modified as follows:

Postulate 2′. A finite line (i.e., segment) can be produced continuously in a line. The line obtained is boundless but not necessarily of infinite extent.

Even with this modification, the axiomatic system consisting of Euclid's first four postulates and the elliptic axiom remains inconsistent since it still yields the following proof of the existence of parallel lines.

Proof of the Existence of Parallel Lines. Let A and B be two points on a line l. Let m and n be lines perpendicular to l at A and B, respectively (11). Assume that m and n are not parallel. Let C be their point of intersection. Then find C' on m on the side of l opposite C such that the segments AC and AC' are congruent (3). Construct $C'B$ (see Fig. 2.43). Then since $\triangle ABC \simeq \triangle ABC'$, and $\angle C'BA \simeq \angle CBA$ (4); it follows that $\angle C'BA$ is also a right angle. Thus, by proposition 14, C', B, and C are collinear, and hence m and n intersect in two distinct points C and C', which yields a contradiction. Thus m and n are parallel lines.   □

So to obtain a consistent axiom system, including the elliptic axiom, the preceding proof must also be invalidated. After some consideration, it should become apparent that the following unstated assumptions were used in the proof:

1. A line separates the plane.
2. Two distinct points lie on a unique line.

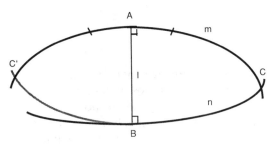

Figure 2.43

Thus this proof would be invalidated if either of these two assumptions were negated. It is the need to negate one of these two assumptions that leads to two types of elliptic geometry. If the first assumption is maintained and the second assumption is negated, that is, if it is assumed that two distinct points do not necessarily lie on a unique line, the geometry known as double elliptic geometry is obtained. If, on the other hand, the second assumption is maintained and the first assumption is negated, that is, if it is assumed that a line does not separate the plane, the geometry known as single elliptic geometry is obtained. Either choice, together with the modification of Postulate 2, results in a system radically different from Euclid's. Hence it is nearly impossible to salvage any of Euclid's work and it becomes essential to develop an entirely new set of axioms for both single and double elliptic geometry. Axioms for these geometries can be found in Chapters 7 and 8 of *An Introduction to Non-Euclidean Geometry* by David Gans (1973). Since models of both double and single elliptic geometry are easily accessible, we can achieve considerable familiarity with these geometries by considering a list of some of the major properties of each geometry. These properties should be apparent within the context of the models to be described here. Thus, we can sample the flavor of elliptic geometries without considering a series of detailed proofs.

Models of Double and Single Elliptic Geometry

Term	Interpretation for Double Elliptic	Interpretation for Single Elliptic
Point	Point on the surface of a Euclidean sphere	Point on the surface of a Euclidean hemisphere if the point is not on the edge; points on the edge are identified with their diametric opposite
Line	Great circle	Semigreat circle
Length	Euclidean length	Euclidean length with the modification implied by the preceding
Angle measure	Euclidean angle measure	Euclidean angle measure

(Diagrams of these models are given in Fig. 2.44 and 2.45.)

In the preceding spherical model, the following properties of double elliptic geometry become apparent.

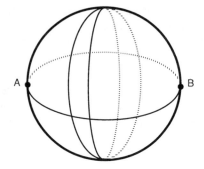

Figure 2.44. Model for double elliptic geometry.

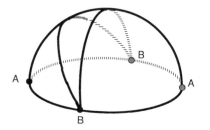

Figure 2.45. Model for single elliptic geometry.

Properties of Double Elliptic Geometry

1. A line separates the plane.
2. There is at least one line through each pair of points.
3. Each pair of lines meets in exactly two points.
4. There is a positive constant k, such that the distance between two points never exceeds πk. Two points at the maximum distance are called *opposite* points.
5. All lines have the same length $2\pi k$.
6. Corresponding to each point there is a unique opposite point.
7. Two points lie on a unique line iff the points are not opposite.
8. All the lines through a given point also pass through the point opposite the given point.
9. All the lines perpendicular to any given line meet in the same pair of opposite points. The distance from each of these points to any point of the given line is $\pi k/2$. These two opposite points are called *poles* of the given line and the line is called the *polar* of the two points.
10. All the lines through a point are perpendicular to the polar of that point.
11. There exists a unique perpendicular to a given line through a given point iff the point is not the pole of the line.

12. The summit angles of a Saccheri quadrilateral are congruent and obtuse.
13. The angle-sum of every triangle exceeds 180°.
14. The area of a triangle is given by area($\triangle ABC$) $= k^2(m(\angle ABC) +$ $m(\angle BCA) + m(CAB) - 180°)$.

Similarly, in the modified hemisphere model the following properties of single elliptic geometry become apparent.

Properties of Single Elliptic Geometry

1. A line does not separate the plane.
2. There is at least one line through each pair of points.
3. Each pair of lines meets in exactly one point.
4. There is a positive constant k, such that the distance between two points never exceeds $\pi k/2$. Two points that divide a line into equal segments are called *opposite* points.
5. On a given line, corresponding to each point there is an opposite point on the line.
6. All lines have the same length πk.
7. All lines perpendicular to any given line go through the same point. The distance from this point to any point of the given line is $\pi k/2$. The point is called the *pole* of the given line and the line is called the *polar* of the point.
8. All the lines through a point are perpendicular to the polar of that point.
9. There exists a unique perpendicular to a given line through a given point iff the point is not the pole of the given line.
10. The summit angles of a Saccheri quadrilateral are congruent and obtuse.
11. The angle sum of every triangle exceeds 180°.
12. The area of a triangle is given by area $(\triangle ABC) = k^2(m(\angle ABC) +$ $m(\angle BCA) + m(\angle CAB) - 180°)$.

EXERCISES

In each of the following exercises, sketch the figure in the spherical model of double elliptic geometry. Assume that the radius of the spherical model is r. Then use your figures to answer the questions asked.

1. (a) Sketch a triangle with three right angles. (b) How long are the sides of this triangle? (c) What fraction of the surface of the sphere does your triangle cover? (d) What is the area of your triangle?

2. (a) Sketch a right triangle with an acute angle. (b) What is the upper bound on the length of the side opposite the acute angle?

3. (a) Sketch a right triangle with an obtuse angle. (b) What is the lower bound on the length of the side opposite the obtuse angle? (c) What is the upper bound on the length of this side?

4. Sketch two triangles that have two pairs of congruent angles and a pair of congruent sides opposite one of the pair of congruent angles. Are these triangles necessarily congruent?

5. Sketch a Saccheri quadrilateral. How does the length of the summit compare with the length of the base?

6. A circle in double elliptic geometry is a set of points at a constant distance (called the *radius*) from a point called the *center*. (a) Sketch circles with radius $\rho < \pi r/2$, $\rho = \pi r/2$, $\pi r/2 < \rho < \pi r$, and $\rho = \pi r$. (Note: In the spherical model the distance is measured along the surface of the sphere.) (b) What other terms can you use to describe the circle with radius $\pi r/2$? With radius πr? (c) Note that a circle with radius ρ and center P can also be described as a circle with radius ρ' and center P'. What is the relationship between ρ and ρ'? Between P and P'?

7. What is the smallest n, for which an ortho-n-gon exists in double elliptic geometry? (An *ortho-n-gon* is a polygon with n sides in which each of the angles is a right angle.)

2.10. Significance of the Discovery of Non-Euclidean Geometries

The development of non-Euclidean geometry began historically with the attempt to prove Euclid's fifth postulate from his first four postulates. By the early 19th century, mathematicians began to accept the possibility that the fifth postulate might be independent. That this postulate is indeed independent was demonstrated when in 1868 the Italian mathematician Eugenio Beltrami (1835–1900) exhibited the first in a series of geometric models of hyperbolic geometry. The best known of these geometric models is the Poincaré model introduced in Section 2.3. Under the interpretations of these models, the axioms of hyperbolic geometry become theorems in Euclidean geometry. Thus hyperbolic geometry was shown to be relatively consistent, and in particular, the models demonstrated that hyperbolic geometry is consistent if Euclidean geometry is. The question of the independence of the fifth postulate had finally been settled!

That the development of non-Euclidean geometry had profound mathematical and philosophical consequences has already been mentioned at the beginning of this chapter. The abstract considerations of these geometries also had important implications in other areas. Riemann's lecture of 1854 used a method that created an infinite number of geometries, and Einstein adopted one of these Riemannian geometries in his study of relativity. A description of this geometry and Einstein's use of it is contained in an essay by Penrose (1978). Furthermore, research since World War II indicates that binocular

Table 2.1. A comparison of Euclidean and Non-Euclidean Geometries*

	Euclidean	Hyperbolic	Elliptic	
Two distinct lines intersect in	at most one	at most one	one (single) two (double)	point(s).
Given a line m and point P not on m, there exist	exactly one	at least two	no lines	through P not intersecting m.
A line	does not	does not	does	have finite length.
A line	is	is	is not	separated by a point.
A line	does	does	does not (single) does (double)	separate the plane.
Nonintersecting lines	are equidistant	are never equidistant	do not exist.	
If a line intersects one of two nonintersecting lines, it	must	may or may not		intersect the other.
The summit angles in a Saccheri quadrilateral are	right	acute	obtuse	angles.
Two distinct lines perpendicular to the same line are	parallel	ultraparallel	intersecting.	
The angle sum of a triangle is	equal to	less than	greater than	180°
The area of a triangle is	independent	proportional to the defect	proportional to the excess	of its angle sum.
Two triangles with congruent corresponding angles are	similar	congruent	congruent.	

*Reprinted with permission from Meserve (1983), *Fundamental Concepts of Geometry*.

visual space is hyperbolic. Descriptions of this research are recorded in Trudeau (1987) and in articles by Ogle (1962) and Zage (1980).

2.11. Suggestions for Further Reading

Aleksandrov, A.D. (1969). Non-Euclidean Geometry. In: A.D. Aleksandrov, A.N. Kolmogorov, and M.A. Lavrent'ev (Eds.), *Mathematics: Its Content, Methods and Meaning*, Vol. 3, pp. 97–189. Cambridge, MA: M.I.T. Press. (This is an expository presentation of non-Euclidean geometry.)

Gans, D. (1973). *An Introduction to Non-Euclidean Geometry*. New York: Academic Press. (This is an easy-to-read and detailed presentation.)

Gray, J. (1979). *Ideas of Space: Euclidean, Non-Euclidean and Relativistic*. Oxford: Clarendon Press.

Heath, T.L. (1956). *The Thirteen Books of Euclid's Elements*, 2d ed. New York: Dover.

Henderson, L.D. (1983). *The Fourth Dimension and Non-Euclidean Geometry in Modern Art*. Princeton, NJ: Princeton University Press.

Lieber, L.R. (1940). *Non-Euclidean Geometry: Or, Three Moons in Mathesis*, 2d ed. New York: Galois Institute of Mathematics and Art. (This is an entertaining poetic presentation.)

Lockwood, J.R., and Runion, G.E. (1978). *Deductive Systems: Finite and Non-Euclidean Geometries*. Reston, VA: N.C.T.M. (This is a brief elementary introduction that can be used as supplementary material at the high-school level.)

Ogle, K.N. (1962). The visual space sense. *Science* 135: 763–771.

Penrose, R. (1978). The geometry of the universe. In: L.A. Steen (Ed.), *Mathematics Today: Twelve Informal Essays*, pp. 83–125. New York: Springer-Verlag.

Ryan, P.J. (1986). *Euclidean and Non-Euclidean Geometry: An Analytic Approach*. Cambridge: Cambridge University Press. (Uses groups and analytic techniques of linear algebra to construct and study models of these geometries.)

Sommerville, D. (1970). *Bibliography of Non-Euclidean Geometry*, 2d ed. New York: Chelsea.

Trudeau, R.J. (1987). *The Non-Euclidean Revolution*. Boston: Birkhauser. (This presentation of both Euclid's original work and non-Euclidean geometry is interwoven with a nontechnical description of the revolution in mathematics that resulted from the development of non-Euclidean geometry.)

Wolfe, H.E. (1945). *Introduction to Non-Euclidean Geometry*. New York: Holt, Rinehart and Winston. (Chap. 1, 2, and 4 contain a development similar to that in this text.)

Zage, W.M. (1980). The geometry of binocular visual space. *Mathematics Magazine* 53(5): 289–294.

Readings on the History of Geometry

Barker, S.F. (1984). Non-Euclidean geometry. In: D.M. Campbell and J.C. Higgins (Eds.), *Mathematics: People, Problems, Results*, Vol. 2, pp. 112–127. Belmont, CA: Wadsworth.

Barker, S.F. (1964). *Philosophy of Mathematics*, pp. 1–55. Englewood Cliffs, NJ: Prentice-Hall.

Bold, B. (1969). *Famous Problems of Geometry and How to Solve Them*. New York: Dover.

Bronowski, J. (1974). The music of the spheres. In: *The Ascent of Man*, pp. 155–187. Boston: Little, Brown.

Eves, H. (1976). *An Introduction to the History of Mathematics*, 4th ed. New York: Holt, Rinehart and Winston.

Gardner, M. (1966). The persistence (and futility) of efforts to trisect the angle. *Scientific American* 214: 116–122.

Gardner, M. (1981). Euclid's parallel postulate and its modern offspring. *Scientific American* 254: 23–24.

Heath, T.L. (1921). *A History of Greek Mathematics*. Oxford: Clarendon Press.

Heath, T.L. (1956). *The Thirteen Books of Euclid's Elements*, 2d ed. New York: Dover.

Hoffer, W. (1975). A magic ratio recurs throughout history. *Smithsonian* 6(9): 110–124.

Kline, M. (1972). *Mathematical Thought from Ancient to Modern Times*, pp. 3–130, 861–881. New York: Oxford University Press.

Knorr, W.R. (1986). *The Ancient Tradition of Geometric Problems*. Boston: Birkhauser.

Maziarz, E., and Greenwood, T. (1984). Greek mathematical philosophy. In: D.M. Campbell and J.C. Higgins (Eds.), *Mathematics: People, Problems, Results*, Vol. 1, pp. 18–27. Belmont, CA: Wadsworth.

Mikami, Y. (1974). *The Development of Mathematics in China and Japan*, 2d ed. New York: Chelsea.

Smith, D.E. (1958). *History of Mathematics*, Vol. 1, pp. 1–147. New York: Dover.

Swetz, F. (1984). The evolution of mathematics in ancient China. In: D.M. Campbell and J.C. Higgins (Eds.), *Mathematics: People, Problems, Results*, Vol. 1, pp. 28–37. Belmont, CA: Wadsworth.

Suggestions for Viewing

A Non-Euclidean Universe (1978; 25 min). Depicts the Poincaré model of the hyperbolic plane. Produced by the Open University Production Centre, Walton Hall, Milton Keynes MK7 6BH, UK.

CHAPTER 3

Geometric Transformations of the Euclidean Plane

3.1. Gaining Perspective

The presentation of non-Euclidean geometry in Chapter 2 was *synthetic*; that is, figures were studied directly and without use of their algebraic representations. This reflects the manner in which both Euclidean and non-Euclidean geometries were orginally developed. However, in the 17th century, French mathematicians Pierre de Fermat (1601–1665) and René Descartes (1596–1650) began using algebraic representations of figures. They realized that by assigning to each point in the plane an ordered pair of real numbers, algebraic techniques could be employed in the study of Euclidean geometry. This study of figures in terms of their algebraic representations by equations is known as *analytic geometry*.

The use of algebraic techniques eventually led to the application of group theory to the study of geometry. This approach led Felix Klein (1849–1925) to give the following definition of geometry in his Erlanger Program of 1872.

Definition 3.1. A *geometry* is the study of those properties of a set S that remain invariant (unchanged) when the elements of S are subjected to the transformations of some transformation group.

Using this definition, Klein was able to give a classification of geometries in terms of groups of linear transformations. The Euclidean transformations are the motions required to carry out the superposition of figures. This technique of moving one figure on top of another to verify congruence is based historically on Euclid's Common Notion 4 and was employed in his proofs of Propositions 4 and 8 (commonly known as the SAS and SSS theorems, respectively).

This transformation approach not only makes Euclidean geometry a more dynamic subject but also introduces the techniques used in contemporary computer graphics. Furthermore, by generalizing the transformations of the Euclidean plane, we will be able to obtain first the transformations of similarity geometry and then those of affine geometry. With a slight change in the set of points, the next step in this generalization yields the transformations of projective geometry. This geometry and its transformations are the subject of Chapter 4.

In this chapter we use the transformation approach to study Euclidean, similarity, and affine geometries. We shall find matrix representations for the appropriate transformations for each geometry and use the techniques of matrix algebra to determine the effects of these transformations. To promote understanding and visualization of the effects of the transformations known as isometries, we will consider symmetry groups of regular polygons and frieze patterns. Interested readers may want to pursue two more delightful topics that will further reinforce ideas from this chapter: tiling the plane and the geometry of paper folding. Sources for these topics are listed at the end of the chapter.

3.2. An Analytic Model of the Euclidean Plane

Before actually describing the analytic model we use, it may be helpful to give some indication of the motivation for choosing this particular model. This discussion will also serve to introduce the terminology and notation that will be used.

The analytic study of Euclidean geometry is based on the premise that each point in the plane can be assigned an ordered pair of real numbers. The usual manner in which this is done is via a Cartesian coordinate system where two perpendicular lines are used as axes. The point of intersection of these axes is assigned the ordered pair $(0, 0)$ and other points are assigned ordered pairs as shown in Fig. 3.1. Rather than denote points by ordered pairs (x, y) as is customary in calculus, we will use ordered pairs (x_1, x_2). This choice is motivated by the "symmetrical form" our results will take in this notation.

With this representation of points, lines of the Euclidean plane can be represented by linear equations of the form $a_1 x_1 + a_2 x_2 + a_3 = 0$ where the a_i are constant real number coefficients. Thus each ordered triple $[a_1, a_2, a_3]$,

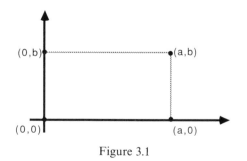

Figure 3.1

where a_1 and a_2 are not both zero determines the equation of a line. Notice that square brackets are used for coordinates of lines so as to distinguish them from coordinates for points. Unlike points, the coordinates of a line do not uniquely represent a line, since the equations $a_1x_1 + a_2x_2 + a_3 = 0$ and $ka_1x_1 + ka_2x_2 + ka_3 = 0$ represent the same line for every nonzero real number k. There is, however, a one-to-one correspondence between the set of lines and the set of *equivalence classes* of ordered triples of real numbers defined by the following relation:

$$[b_1, b_2, b_3] \sim [a_1, a_2, a_3] \quad \text{if } b_i = ka_i, \ i = 1, 2, 3, \text{ where}$$
$$k \text{ is a nonzero real number}$$

Using Definition 3.2, we can show that this relation is an equivalence relation (see Exercise 8).

Definition 3.2. A relation "$\sim$" is an *equivalence relation* if its satisfies each of the following:

(a) $a \sim a$.
(b) If $a \sim b$ then $b \sim a$.
(c) If $a \sim b$ and $b \sim c$, then $a \sim c$.

Definition 3.3. A set of elements, all of which are pairwise related by an equivalence relation is called an *equivalence class*. Any element of an equivalence class is called a *representative* of the equivalence class.

Since there is a one-to-one correspondence between the lines of the Euclidean plane and these equivalence classes, we can interpret lines in terms of these equivalence classes. The ordered triples $[u_1, u_2, u_3]$ belonging to a particular equivalence class will be called *homogeneous coordinates* of the line. If we consider one of these ordered triples to be a row matrix $u[u_1, u_2, u_3]$, then the equation of the corresponding line is $uX = 0$ where $X = (x_1, x_2, 1)$ is a column matrix with 1 in its third entry. In particular if $u = [2, -3, 5]$, then $uX = 0$ is the equation $2x_1 - 3x_2 + 5 = 0$. This observation, together with the desire to use similar interpretations for points and lines, suggests that we interpret points in terms of equivalence classes of ordered triples of real numbers (x_1, x_2, x_3), where $x_3 \neq 0$ under the same relation. Again we will refer to elements of these equivalence classes as *homogeneous coordinates* of the point. In the case of points, however, since x_3 is always nonzero, every ordered triple $(x_1, x_2, x_3) \sim (x_1/x_3, x_2/x_3, 1)$ so each equivalence class will have a unique representative of the form $(x_1, x_2, 1)$. In other words, each point in the plane that we are accustomed to denoting with an ordered pair of the form (x_1, x_2) can now be denoted by the corresponding ordered triple $(x_1, x_2, 1)$. For example, instead of referring to a point with coordinates $(1, -3)$ we will now refer to it as a point with coordinates $(1, -3, 1)$.

Analytic Model for the Euclidean Plane

Undefined Term	Interpretation
Points	Equivalence classes of ordered triples (x_1, x_2, x_3) where $x_3 \neq 0$ (any one of the representatives of an equivalence class will be called coordinates of the point)
Lines	Equivalence classes of ordered triples $[u_1, u_2, u_3]$ where u_1 and u_2 are not both 0 (any one of the representatives of the equivalence class will be called coordinates of the line)
Incidence	A point $X(x_1, x_2, x_3)$ is incident with a line $u[u_1, u_2, u_3]$ iff $[u_1, u_2, u_3] \cdot (x_1, x_2, x_3) = 0$ or in matrix notation $uX = 0$

As indicated before, lines will always be represented by row matrices and points by column matrices. But unlike the conventional algebra usage of capital letters for all matrices, here matrices of line coordinates will be represented by lowercase letters.

Within the context of this analytic model, the operations of matrix algebra take on geometric significance as indicated by the following theorems. In each case the coordinates chosen to represent points will be those in which $x_3 = 1$. The first of these theorems gives a convenient way to determine when three points are *collinear*, that is, on the same line.

Theorem 3.1. *Three distinct points* $X(x_1, x_2, 1)$, $Y(y_1, y_2, 1)$, *and* $Z(z_1, z_2, 1)$ *are collinear iff the determinant*

$$\begin{vmatrix} x_1 & y_1 & z_1 \\ x_2 & y_2 & z_2 \\ 1 & 1 & 1 \end{vmatrix} = 0$$

Proof. X, Y, Z are collinear iff there is a line $u[u_1, u_2, u_3]$ such that

$$u_1 x_1 + u_2 x_2 + u_3 = 0$$
$$u_1 y_1 + u_2 y_2 + u_3 = 0$$
$$u_1 z_1 + u_2 z_2 + u_3 = 0$$

or

$$[u_1, u_2, u_3] \begin{bmatrix} x_1 & y_1 & z_1 \\ x_2 & y_2 & z_2 \\ 1 & 1 & 1 \end{bmatrix} = [0, 0, 0]$$

But from linear algebra this equation has a nontrivial solution $[u_1, u_2, u_3]$ iff

$$\begin{vmatrix} x_1 & y_1 & z_1 \\ x_2 & y_2 & z_2 \\ 1 & 1 & 1 \end{vmatrix} = 0$$

Since this nontrivial solution cannot have both $u_1 = 0$ and $u_2 = 0$ (see Exercise 7), $u[u_1, u_2, u_3]$ is a line containing all three points.   □

Corollary. *If A and B are distinct points, then the equation of the line AB where* $A(a_1, a_2, 1)$ *and* $B(b_1, b_2, 1)$ *can be written*

$$\begin{vmatrix} x_1 & a_1 & b_1 \\ x_2 & a_2 & b_2 \\ 1 & 1 & 1 \end{vmatrix} = 0$$

In the proof of Theorem 3.1 we used the familiar notion that the equation $u_1 x_1 + u_2 x_2 + u_3 = 0$ determines which points lie on the line with coordinates $[u_1, u_2, u_3]$, and we refer to this equation as the *equation of the line u*. We often think of the values of the u_i as constants. For example, the equation $3x_1 - 4x_2 + 10 = 0$ determines which points lie on the line with coordinates $[3, -4, 10]$. However, it is equally useful to regard $u_1 x_1 + u_2 x_2 + u_3 = 0$ as the *equation of the point X* and use it to determine which lines pass through the point with coordinates $(x_1, x_2, 1)$. In particular, we can determine which lines pass through the point with coordinates $(-2, 5, 1)$ by finding ordered triples $[u_1, u_2, u_3]$ that satisfy the equation $-2u_1 + 5u_2 + u_3 = 0$.

With the previous discussion in mind, we can use line coordinates to determine when three lines are *concurrent*, that is, when all three lines intersect at a common point. The proof of this theorem is similar to the previous theorem except here special consideration needs to be given to the case where the only nontrivial solutions are those for which $x_3 = 0$ (see Exercise 12).

Theorem 3.2. *Three distinct lines u, v, w are all concurrent or all parallel iff the determinant*

$$\begin{vmatrix} u_1 & u_2 & u_3 \\ v_1 & v_2 & v_3 \\ w_1 & w_2 & w_3 \end{vmatrix} = 0$$

Corollary. *The equation of the point of intersection of concurrent lines p and q, denoted p·q, can be written*

$$\begin{vmatrix} u_1 & u_2 & u_3 \\ p_1 & p_2 & p_3 \\ q_1 & q_2 & q_3 \end{vmatrix} = 0$$

In these theorems, it is important to note that the coordinates of points appear in columns whereas the coordinates of lines appear in rows. This convention is used throughout the remainder of the text.

The line coordinates can also be used to determine the angle between two lines using a definition given in terms of the tangent of the angle. This definition makes use of a formula from trigonometry which gives the tangent of the angle between two lines in terms of the slopes of the lines (see Exercise 17).

Definition 3.4. If $u[u_1, u_2, u_3]$ and $v[v_1, v_2, v_3]$ are two lines, then the *angle between u and v*, denoted $\angle(u, v)$, is defined to be the unique angle such that

$$\tan(\angle(u, v)) = \frac{u_1 v_2 - u_2 v_1}{u_1 v_1 + u_2 v_2} \quad \text{and} \quad \begin{array}{l} -90° < m(\angle(u, v)) < 90° \\ \text{if } u_1 v_1 + u_2 v_2 \neq 0 \end{array}$$

$$m(\angle(u, v)) = 90° \qquad\qquad\qquad \text{if } u_1 v_1 + u_2 v_2 = 0$$

Note that this definition is independent of the particular set of homogeneous coordinates used for the lines and that only the first two coordinates of each line are used. This corresponds to defining the angle between lines u and v in terms of the angle between lines u' and v' where the latter are lines through the point $(0, 0, 1)$ and u' is parallel to u, v' is parallel to v (see Exercise 13). In particular, this definition assigns as the angle between parallel lines the angle with measure 0.

EXERCISES

1. Let u be the line with homogeneous coordinates $[-2, 5, 7]$. (a) Find three other sets of coordinates for u. (b) Find an equation for line u. (c) Find coordinates for two distinct points on u.

2. Let P be the point with ordered pair coordinates $(4, -7)$. (a) Find three sets of homogeneous coordinates for P. (b) Find an equation for the point P. (c) Find coordinates for two lines through P.

3. Find homogeneous coordinates for each of the following: (a) the x_1 axis; (b) the x_2 axis; and (c) the line $x_1 = x_2$.

4. Find the general form of the coordinates for lines through the point $(0, 0, 1)$.

5. Use the corollary to Theorem 3.1 to find the line containing the points $(10, 2)$ and $(-7, 3)$.

6. Use the corollary to Theorem 3.2 to find the point of intersection of the lines $3x + 4y + 7 = 0$ and $2x - y + 8 = 0$.

7. Show that the nontrivial solution obtained in the proof of Theorem 3.1 cannot have both $u_1 = 0$ and $u_2 = 0$.

8. Show that the following is an equivalence relation: $[u_1, u_2, u_3] \sim [v_1, v_2, v_3]$ if $u_i = k v_i$ for some nonzero k.

9. Prove the corollary to Theorem 3.1.

10. Show algebraically that two distinct lines $u[u_1, u_2, u_3]$ and $v[v_1, v_2, v_3]$ are parallel (do not intersect) iff $u_1 = k v_1$, $u_2 = k v_2$, but $u_3 \neq k v_3$ for some nonzero real number k. (*Hint*: Show that the system of equations

$u_1 x_1 + u_2 x_2 + u_3 = 0$ and $v_1 x_1 + v_2 x_2 + v_3 = 0$ does *not* have a solution iff these conditions are true.)

11. Use the result of Exercise 10 to verify that Playfair's axiom is true in this analytic model of the Euclidean plane.

12. Use the result of Exercise 7 to prove Theorem 3.2. (Be sure to see the comment preceding the theorem.)

13. Use the result of Exercise 10 to show that the line parallel to $u[u_1, u_2, u_3]$, which passes through the point $(0, 0, 1)$, has coordinates $[u_1, u_2, 0]$.

14. Using Definition 3.4, find the angles between the following lines: (a) the lines $[-2, 1, 7]$ and $[3, 4, 17]$; (b) the x_1 and x_2 axes; and (c) the line $x_1 = x_2$ and the x_1 axis.

15. Use the result of Exercise 10 to find the angle between two parallel lines.

16. Let the line u be the x_1 axis and the line v be the line with coordinates $[v_1, v_2, v_3]$. Use Definition 3.4 to show that $\tan(\angle(u, v)) = -(v_1)/(v_2)$. [Recall that the *slope* of the line $[v_1, v_2, v_3]$ is given by $-(v_1)/(v_2)$.]

17. The following trigonometry formula gives the tangent of the angle between lines u and v in terms of their slopes m_u and m_v. Use the definition of slope from Exercise 16 to show that this formula is equivalent to the formula used in Definition 3.4:

$$\tan(\angle(u, v)) = \frac{m_v - m_u}{1 + m_u m_v}$$

18. Prove: If P is a point and l is a line, there is a unique line through P perpendicular to l.

3.3. Linear Transformations of the Euclidean Plane

The transformation approach to the study of Euclidean geometry involves identification of appropriate groups of transformations of the Euclidean plane and the investigation of the features preserved by these groups. In this section we introduce the definitions and theorems from linear algebra needed to pursue this approach. Since the analytic model of the Euclidean plane interprets points and lines in terms of equivalence classes of the vector space R^3, we use a special set of functions whose domain and range are both R^3.

Definition 3.5. Let V be a vector space over R. If $T: V \to V$ is a function, then T is called a *linear transformation of V* if it satisfies both the following conditions: (1) $T(\mathbf{u} + \mathbf{v}) = T(\mathbf{u}) + T(\mathbf{v})$ for all vectors $\mathbf{u}$ and $\mathbf{v}$ in V; and (2) $T(k\mathbf{u}) = kT(\mathbf{u})$ for all vectors $\mathbf{u}$ in V and scalars k in R.

Definition 3.6. A linear transformation T is *one-to-one* if whenever $\mathbf{u} \neq \mathbf{v}$, $T(\mathbf{u}) \neq T(\mathbf{v})$.

From these definitions it should be clear that the equivalence classes of R^3 defined by the relation $\mathbf{u} \sim \mathbf{v}$ iff $\mathbf{u} = k\mathbf{v}$ are preserved by linear transformations. In other words, if $\mathbf{u} \sim \mathbf{v}$, then $T(\mathbf{u}) \sim T(\mathbf{v})$. So a one-to-one linear transformation of R^3 induces a one-to-one mapping on the set of points of the model for the Euclidean plane. Each of these mappings has a *matrix representation* as indicated by the following summary of results from linear algebra.

Theorem 3.3. *T is a one-to-one linear transformation of* $R^3 = \{X(x_1, x_2, x_3): x_i \in R\}$ *iff* $T(X) = AX$ *where* $A = [a_{ij}]_{3 \times 3}$, $|A| \neq 0$ *and* $a_{ij} \in R$.

Because we make use of homogeneous coordinates of the form $(x_1, x_2, 1)$ for points, it is important to note the following corollary, which can be verified using matrix multiplication (see Exercise 6).

Corollary. *T is a one-to-one linear transformation of* $V^* = \{X(x_1, x_2, 1): x_i \in R\}$ *iff* $T(X) = AX$ *where*

$$A = \begin{bmatrix} a_{11} & a_{12} & a_{13} \\ a_{21} & a_{22} & a_{23} \\ 0 & 0 & 1 \end{bmatrix} \quad |A| \neq 0 \quad \text{and} \quad a_{ij} \in R$$

As indicated previously, we need to verify that the set of one-to-one linear transformations of V^* form a group under composition. If T_1 and T_2 are transformations of a vector space V, the *composite* (or *product*) $T_1 T_2$ is the mapping defined by $(T_1 T_2)(\mathbf{u}) = T_1(T_2(\mathbf{u}))$ for all vectors $\mathbf{u}$ in V. Because the composition of functions is *associative*, that is, $T_1(T_2 T_3) = (T_1 T_2) T_3$, we can use the following simplified definition.

Definition 3.7. A nonempty set G of transformations of a vector space V is said to form a *group* under the operation of composition if it satisfies both the following conditions: (1) If $T \in G$ then $T^{-1} \in G$; and (2) if $T_1 \in G$ and $T_2 \in G$, then $T_1 T_2 \in G$.

Note that the definition guarantees that any group G contains a transformation T, and therefore by property 1, T^{-1} is in G. Property 2 then tells us that $TT^{-1} = I$ is also in G where I is the *identity* transformation defined by $I(\mathbf{u}) = \mathbf{u}$ for any $\mathbf{u}$ in V.

An application of results of matrix algebra to the matrix representations given by the corollary to Theorem 3.3 can be used to prove the following theorem (see Exercise 9).

Theorem 3.4. *The set of one-to-one linear transformations of* V^* *is a group.*

Even though each of the transformations of this group can be represented by a 3×3 matrix with real number entries and the images of individual points can be computed algebraically, it is important to visualize the geometric action of each transformation as a mapping or moving of all of the points of the Euclidean plane to other points of the plane. Determining the general way in which a transformation moves points, and, in particular, determining which points and lines it leaves unaffected, are essential to understanding this geometric action.

EXAMPLE 3.1. Let T be the linear transformation with matrix A shown. If X is a point on the line $l[1, -1, 0]$, show that $T(X)$ is also a point on l and show that $T(P) = P$ where P is the point $P(-\frac{2}{3}, -\frac{2}{3}, 1)$.

$$A = \begin{bmatrix} 1 & 3 & 2 \\ 3 & 1 & 2 \\ 0 & 0 & 1 \end{bmatrix}$$

To find the images of points on l, we note that $X(x_1, x_2, 1)$ is on l iff $x_1 - x_2 = 0$, that is, if $x_2 = x_1$. Thus we can find the images of any point X on l as follows:

$$\begin{bmatrix} 1 & 3 & 2 \\ 3 & 1 & 2 \\ 0 & 0 & 1 \end{bmatrix} \begin{bmatrix} x_1 \\ x_1 \\ 1 \end{bmatrix} = \begin{bmatrix} 4x_1 + 2 \\ 4x_1 + 2 \\ 1 \end{bmatrix}$$

Since $x_1' = 4x_1 + 2 = x_2'$, it is clear that $T(X) = X'(x_1', x_2', 1)$ is also a point on l. Furthermore, since P is a point on l, we can set $x_1 = x_2 = -\frac{2}{3}$ in the preceding computation to find $T(P)$. With this value of x_1, we get $x_1' = x_2' = 4(-\frac{2}{3}) + 2 = -\frac{2}{3}$. So $T(P) = P$.

Since the image of P is itself under the transformation T in this example we say that P is an *invariant point* of the transformation. Furthermore, since the images of all points on l are again on l, we say that l is an *invariant line* of T. Note, however, that the points on l other than P are not invariant, so l is not *pointwise invariant*.

Definition 3.8. A property that is unchanged under a transformation is called an *invariant* of the transformation. A property that is invariant under each transformation of a group of transformations is called an *invariant* of the group. An invariant property of a transformation is said to be *preserved* by the transformation.

Essential to the study of any mathematical system is a determination of the transformations that preserve certain features of the system. The following result shows that the group of one-to-one linear transformations of V^* preserve collinearity; that is, collinearity is one of the invariant properties of this group.

Thus these transformations of the points of the Euclidean plane also map lines to lines. We say they *induce* mappings between lines of the Euclidean plane.

Theorem 3.5. *A one-to-one linear transformation of V^* preserves collinearity (i.e., the images of collinear points are collinear).*

Proof. Let $X'(x'_1, x'_2, 1)$, $Y'(y'_1, y'_2, 1)$, and $Z'(z'_1, z'_2, 1)$ be images of the points X, Y, and Z under a given one-to-one linear transformation with matrix A. Then combining all our efforts into one matrix equation, we get

$$\begin{bmatrix} x'_1 & y'_1 & z'_1 \\ x'_2 & y'_2 & z'_2 \\ 1 & 1 & 1 \end{bmatrix} = A \begin{bmatrix} x_1 & y_1 & z_1 \\ x_2 & y_2 & z_2 \\ 1 & 1 & 1 \end{bmatrix}$$

and taking determinants of both sides yields

$$\begin{vmatrix} x'_1 & y'_1 & z'_1 \\ x'_2 & y'_2 & z'_2 \\ 1 & 1 & 1 \end{vmatrix} = |A| \begin{vmatrix} x_1 & y_1 & z_1 \\ x_2 & y_2 & z_2 \\ 1 & 1 & 1 \end{vmatrix}$$

Therefore, the result follows by Theorem 3.1. □

Note that this theorem also implies that one-to-one linear transformations preserve incidence. In other words, if the point X is on line u, then X', the image of X, is on u', the image of u.

Just as the image of a point under a one-to-one linear transformation can be determined by a matrix equation, a matrix equation can be used to determine the image of a line under the *same* transformation. This second equation is related to but *not* identical to the first.

Theorem 3.6. *If the image of a point under a one-to-one linear transformation of V^* is given by the matrix equation $X' = AX$ then the image of a line under this same transformation is given by the matrix equation $ku' = uA^{-1}$ for some nonzero scalar k.*

Proof. Consider the line $u[u_1, u_2, u_3]$ with equation $u_1 x_1 + u_2 x_2 + u_3 x_3 = 0$; that is, $uX = 0$. Under the linear transformation, u maps to u', X maps to X', and $uX = 0$ iff $u'X' = 0$. But $X' = AX$. So substituting, $u'AX = 0$ iff $uX = 0$. Since this must hold for all points X, $u = ku'A$ for a nonzero scalar k, or $ku' = uA^{-1}$. □

The transformation with matrix A is said to have *point equation* $X' = AX$ and *line equation* $ku' = uA^{-1}$. While the use of the point equation is straightforward, the scalar k in the line equation (required because there is no unique set of homogeneous coordinates for a given line) makes the use of the line equation slightly more difficult. It is important to note that k is *not* constant for a given matrix A. This becomes especially significant when a given line needs to be mapped to a particular line, as in the following example.

EXAMPLE 3.2. Find the matrix of a one-to-one linear transformation of the Euclidean plane that maps $u[1, -3, 2]$ to $u'[1, 0, -4]$, $v[2, 1, -5]$ to $v'[10, -7, 7]$, and $w[1, -2, 0]$ to $w'[0, 1, -6]$.

Since we are told the images of three lines, we will begin with the general form of the line equation of a linear transformation of V^*, $ku' = uA^{-1}$. Each line and its image will give a matrix equation, as shown. In these equations we let $B = A^{-1}$ and use three distinct values of k:

$$k_1[1, 0, -4] = [1, -3, 2] \begin{bmatrix} b_{11} & b_{12} & b_{13} \\ b_{21} & b_{22} & b_{23} \\ 0 & 0 & 1 \end{bmatrix}$$

$$k_2[10, -7, 7] = [2, 1, -5] \begin{bmatrix} b_{11} & b_{12} & b_{13} \\ b_{21} & b_{22} & b_{23} \\ 0 & 0 & 1 \end{bmatrix}$$

$$k_3[0, 1, -6] = [1, -2, 0] \begin{bmatrix} b_{11} & b_{12} & b_{13} \\ b_{21} & b_{22} & b_{23} \\ 0 & 0 & 1 \end{bmatrix}$$

The resulting system of nine equations in nine unknowns yields the following solutions for the values of k: $k_1 = -2$, $k_2 = 1$, and $k_3 = -1$. The matrix

$$A^{-1} = \begin{bmatrix} 4 & -3 & 6 \\ 2 & -1 & 0 \\ 0 & 0 & 1 \end{bmatrix} \quad \text{and} \quad A = \begin{bmatrix} -\frac{1}{2} & \frac{3}{2} & 3 \\ -1 & 2 & 6 \\ 0 & 0 & 1 \end{bmatrix}$$

EXERCISES

1. Let T be the transformation with matrix

$$A = \begin{bmatrix} 1 & 5 & 0 \\ 0 & 1 & 0 \\ 0 & 0 & 1 \end{bmatrix}$$

(a) Using the technique of Example 3.1, find the images of points on the line $l[1, -2, 3]$. (b) Does T keep any points on l invariant? If so, which one(s)? (c) Use the coordinates of the images of two points on l to find the coordinates of $l' = T(l)$. (d) Sketch both l and l' in the Euclidean plane and describe the geometric action of T.

2. Verify by calculation that the transformation with the matrix in Example 3.2 does indeed map the three lines as desired.

3. Let

$$A = \begin{bmatrix} 1 & 3 & -7 \\ 2 & 5 & 4 \\ 0 & 0 & 1 \end{bmatrix}$$

be the matrix of a transformation, T. (a) Find $P' = T(P)$ and $Q' = T(Q)$ for the points $P(1, 2, 1)$ and $Q(6, 4, 1)$. (b) Find coordinates of the lines PQ and $P'Q'$. (c) Find the matrix of A^{-1} to be used in the line equation of the transformation T. (d) Use this line equation to find the image of the line PQ under T. (Your answer should be the line $P'Q'$.)

4. Find the matrix of a linear transformation that maps $P(0, 0, 1)$ to $P'(1, 5, 1)$, $Q(1, 3, 1)$ to $Q'(3, -7, 1)$, and $R((1, 0, 1)$ to $R'(3, 6, 1)$.

5. Find the matrix of a linear transformation that maps $u[2, -3, 1]$ to $u'[2, 5, 0]$, $v[1, -2, 0]$ to $v'[1, 1, -6]$, and $w[1, 0, 0]$ to $w'[3, 2, -1]$.

6. Prove the corollary of Theorem 3.3.

7. Prove: If A is a matrix of the form given by the corollary of Theorem 3.3, then A^{-1} is also of this form. (Hint: Since these are 3×3 matrices with a third row of the form $(0, 0, 1)$ the adjoint method provides an easy way of computing A^{-1}.)

8. Prove: If A and B are matrices of the form described by the corollary to Theorem 3.3, then the matrix product AB is also of this form.

9. Using results of Exercises 7 and 8, prove Theorem 3.4.

10. Find examples of matrices of one-to-one transformations of V^* such that $AB \neq BA$. (This example shows that this group does not have the *commutative* property.)

3.4. Isometries

To begin our transformation approach to the study of Euclidean geometry we will single out the subset of one-to-one linear transformations of V^* that preserve distance.

Definition 3.9. The *(Euclidean) distance* between two points $X(x_1, x_2, 1)$ and $Y(y_1, y_2, 1)$ is given by $d(X, Y) = \sqrt{(x_1 - y_1)^2 + (x_2 - y_2)^2}$.

Definition 3.10. A one-to-one linear transformation of V^* onto itself is an *isometry* if it preserves distance (i.e., if $d(X, Y) = d(T(X), T(Y))$ for all pairs of points X, Y).

As one-to-one linear transformations of V^*, isometries can be represented by matrices of the form given in the corollary to Theorem 3.3. However, the distance preserving property further restricts the form of their matrix representation.

Theorem 3.7. *An isometry has one of the following matrix representations:*

$$\begin{bmatrix} a_{11} & a_{12} & a_{13} \\ -a_{12} & a_{11} & a_{23} \\ 0 & 0 & 1 \end{bmatrix} \quad \text{or} \quad \begin{bmatrix} a_{11} & a_{12} & a_{13} \\ a_{12} & -a_{11} & a_{23} \\ 0 & 0 & 1 \end{bmatrix}$$

where $(a_{11})^2 + (a_{12})^2 = 1.$

Proof. Let $X'(x'_1, x'_2, 1)$, $Y'(y'_1, y'_2, 1)$ be the images of the points $X(x_1, x_2, 1)$ and $Y(y_1, y_2, 1)$ under an isometry. Then by the corollary to Theorem 3.3:

$$\begin{bmatrix} x'_1 \\ x'_2 \\ 1 \end{bmatrix} = \begin{bmatrix} a_{11} & a_{12} & a_{13} \\ a_{21} & a_{22} & a_{23} \\ 0 & 0 & 1 \end{bmatrix} \begin{bmatrix} x_1 \\ x_2 \\ 1 \end{bmatrix} = \begin{bmatrix} a_{11}x_1 + a_{12}x_2 + a_{13} \\ a_{21}x_1 + a_{22}x_2 + a_{23} \\ 1 \end{bmatrix}$$

and likewise

$$\begin{bmatrix} y'_1 \\ y'_2 \\ 1 \end{bmatrix} = \begin{bmatrix} a_{11}y_1 + a_{12}y_2 + a_{13} \\ a_{21}y_1 + a_{22}y_2 + a_{23} \\ 1 \end{bmatrix}$$

Because this mapping is an isometry $d(X', Y') = d(X, Y)$. From the definition of distance this equality yields

$$[(x_1 - y_1)^2 + (x_2 - y_2)^2]^{1/2}$$
$$= [(x'_1 - y'_1)^2 + (x'_2 - y'_2)^2]^{1/2}$$
$$= [(a_{11}x_1 + a_{12}x_2 - a_{11}y_1 - a_{12}y_2)^2 + (a_{21}x_1 + a_{22}x_2 - a_{21}y_1 - a_{22}y_2)^2]^{1/2}$$
$$= [(a_{11}^2 + a_{21}^2)(x_1 - y_1)^2 + 2(a_{11}a_{12} + a_{21}a_{22})(x_1 - y_1)(x_2 - y_2)$$
$$+ (a_{12}^2 + a_{22}^2)(x_2 - y_2)^2]^{1/2}$$

Since this equality must hold for all points X and Y and therefore for all ordered pairs of real numbers (x_1, x_2) and (y_1, y_2), we can square the first and last of these expressions and then equate the coefficients of like terms. This gives the following equations:

(a) $a_{11}^2 + a_{21}^2 = 1$

(b) $a_{12}^2 + a_{22}^2 = 1$

(c) $a_{11}a_{12} + a_{21}a_{22} = 0$

(d) $a_{11}a_{12} = -a_{21}a_{22}$

To solve these equations, we consider two cases.
Case 1. $a_{11} \neq 0$. Equation (d) implies $a_{12} = -(a_{21}a_{22})/a_{11}$. Substituting this

into (b) gives

$$\frac{a_{21}^2 a_{22}^2}{a_{11}^2} + a_{22}^2 = 1 \quad \text{or} \quad (a_{11}^2 + a_{21}^2)a_{22}^2 = a_{11}^2.$$

But by (a) this becomes $a_{22} = \pm a_{11}$.
If $a_{22} = a_{11}$ then Equation (d) implies $a_{21} = -a_{12}$.
If $a_{22} = -a_{11}$ then Equation (d) implies $a_{21} = a_{12}$.
These results yield the two forms of the matrix given earlier.
    Case 2. $a_{11} = 0$. Again in this case the equations yield the same two forms of the matrix (see Exercise 3).   □

    The determinant of the first isometry matrix is $(a_{11})^2 + (a_{12})^2 = 1$, while the determinant of the second matrix is $-(a_{11})^2 - (a_{12})^2 = -1$. This observation gives a convenient way to distinguish the two types of isometries.

Definition 3.11. If the determinant of the matrix of an isometry is $+1$, the isometry is said to be a *direct isometry*. If the determinant is -1, the isometry is said to be an *indirect isometry*.

    A relatively straightforward argument demonstrates that the isometries form a group. However, since there are direct and indirect isometries, it is necessary to verify that the inverse of each type is an isometry. It is also necessary to demonstrate that the composite $T_1 T_2$ is an isometry where T_1 and T_2 are both direct, both indirect, direct and indirect, and finally indirect and direct isometries, respectively. The results of these computations are summarized in the following theorem and corollary.

Theorem 3.8. *The set of isometries forms a group, of which the set of direct isometries is a subgroup.*

Corollary. *The product of two direct or two indirect isometries is a direct isometry. The product of a direct and an indirect isometry in either order is an indirect isometry.*

    Having identified the set of isometries as a group of transformations, we can study Euclidean geometry by determining which properties of sets of points in V^* are preserved by this group. The following definition formalizes Euclid's use of "superposition."

Definition 3.12. Two sets of points α and β are *congruent*, denoted $\alpha \simeq \beta$, if β is the image of α under an isometry.

    Two specific sets of points whose congruence is invariably studied in any presentation of Euclidean geometry are those figures known as segments and triangles. Before considering these particular figures, it is necessary to accept the following definitions.

Definition 3.13. P is *between* Q and R if P, Q, and R are three distinct collinear points and $d(Q, P) + d(P, R) = d(Q, R)$. The set of points consisting of Q and R together with all points P between Q and R is called the *segment* with *endpoints* Q and R and is denoted $\overline{QR}$. The *measure* of $\overline{QR}$, denoted by $m(\overline{QR})$, is $d(Q, R)$.

Definition 3.14. If P, Q, and R are three noncollinear points then triangle PQR, denoted by $\triangle PQR$, is the set of segments $\overline{PQ}, \overline{QR}$, and $\overline{RP}$. These segments are called the *sides* of the triangle. $\angle PQR$, $\angle QRP$, and $\angle RPQ$ are called the *angles* of the triangle.

Using Definition 3.13 and the definition of an isometry it is relatively easy to verify that congruent segments have the same measure (see Exercise 5).

Theorem 3.9. *If* $\overline{PQ} \simeq \overline{P'Q'}$, *then* $m(\overline{PQ}) = m(\overline{P'Q'})$.

In the case of congruent triangles, we are interested in knowing not only if the corresponding sides of the triangles have the same measure but also how the measures of corresponding angles compare. As the following theorem indicates, the angle measure is unchanged under a direct isometry, but under an indirect isometry the sign of the angle measure is changed. For this reason, we say that indirect isometries *reverse orientation*.

Theorem 3.10. *Let* u' *and* v' *be the images of lines* u *and* v *under an isometry. If the isometry is direct then* $m(\angle (u', v')) = m(\angle (u, v))$. *If the isometry is indirect then* $m(\angle (u', v')) = - m(\angle (u, v))$.

Proof. Let $u[u_1, u_2, u_3]$ and $v[v_1, v_2, v_3]$ be two lines and $u'[u'_1, u'_2, u'_3]$ and $v'[v'_1, v'_2, v'_3]$ their images. The theorem will be proved by showing that $\tan(\angle (u', v')) = \pm \tan(\angle (u, v))$. Using the line equation of an isometry we have $k_1 u' = uA^{-1}$, $k_2 v' = vA^{-1}$ where k_1 and k_2 are nonzero scalars and A is the matrix of an isometry. If we let $B = A^{-1}$, then B is also the matrix of an isometry since the isometries form a group. Thus matrix B has one of the two forms given in Theorem 3.7. In both cases the coordinates of u' and v' can be determined by direct calculation and substituted into the expression for $\tan(\angle (u', v'))$. Simplification of the resulting expressions yields the results given in the theorem (see Exercise 7). $\qquad\square$

The previous two theorems lead to the following result.

Theorem 3.11. *If* $\triangle PQR \simeq \triangle P'Q'R'$ *then* $m(\overline{PQ}) = m(\overline{P'Q'})$, $m(\overline{QR}) = m(\overline{Q'R'})$, $m(\overline{RP}) = m(\overline{R'P'})$, $m(\angle PQR) = \pm m(\angle P'Q'R')$, $m(QRP) = \pm(\angle Q'R'P')$, *and* $m(\angle RPQ) = \pm m(\angle R'P'Q')$.

To show that the converse of this theorem is also valid, it is most convenient to first investigate and classify the isometries. This will be done in the next two sections.

EXERCISES

1. Find both a direct isometry and an indirect isometry that map $X(0, 0, 1)$ and $Y(2, 0, 1)$ to $X'(1, 1, 1)$ and $Y'(3, 1, 1)$. What happens to the point $Z(1, -1, 1)$ under each of these isometries?

2. Prove that the distance function given in Definition 3.9 satisfies properties (a)–(c) in the following definition. (It is also possible to verify that the distance function satisfies property (d) and is therefore a metric on V^*.)

Definition. A function $d(P, Q)$ on a set S is a *metric* if for all points P, Q and R in S: (a) $d(P, Q)$ is a real number; (b) $d(P, Q) = d(Q, P)$; (c) $d(P, Q) \geqslant 0$ and $d(P, Q) = 0$ iff $P = Q$; (d) $d(P, R) \leqslant d(P, Q) + d(Q, R)$.

3. Prove case 2 of Theorem 3.7.

4. Prove Theorem 3.8 and its corollary. (Note: $G' \subset G$ is a subgroup of G if G' is also a group.)

5. Prove Theorem 3.9.

6. Show that the congruence relation as given in Definition 3.12 is an equivalence relation (see Definition 3.2).

7. Carry out the algebraic computations needed to complete the proof of Theorem 3.10 for both direct and indirect isometries.

8. Prove: If $\alpha \simeq \beta$ and $T(\alpha) = \alpha'$, $T(\beta) = \beta'$ where T is an isometry, then $\alpha' \simeq \beta'$.

3.5. Direct Isometries

In Section 3.4, we described isometries as being either direct or indirect. In this section we will investigate and further classify the direct isometries based on the number of points that remain invariant under the isometry. Knowing which points and lines are invariant is important in understanding the geometric action of any transformation.

Theorem 3.12. *A direct isometry other than the identity, with matrix* $A = [a_{ij}]$ *has exactly one invariant point iff* $a_{11} \neq 1$.

Proof. The point $X(x_1 x_2, 1)$ is an invariant point of the isometry iff $AX = X$;

$$\begin{bmatrix} a_{11} & a_{12} & a_{13} \\ -a_{12} & a_{11} & a_{23} \\ 0 & 0 & 1 \end{bmatrix} \begin{bmatrix} x_1 \\ x_2 \\ 1 \end{bmatrix} = \begin{bmatrix} x_1 \\ x_2 \\ 1 \end{bmatrix}$$

or

$$(a_{11} - 1)x_1 + a_{12}x_2 + a_{13} = 0 \qquad (3.1)$$

and

$$-a_{12}x_1 + (a_{11} - 1)x_2 + a_{23} = 0 \qquad (3.2)$$

Case 1. $a_{11} \neq 1$. In this case, Equation (3.1) yields

$$x_1 = \frac{-a_{12}x_2 - a_{13}}{a_{11} - 1}$$

and Equation (3.2) yields

$$-a_{12} \left[\frac{-a_{12}x_2 - a_{13}}{a_{11} - 1} \right] + (a_{11} - 1)x_2 + a_{23} = 0$$

or solving for x_2

$$x_2 = \frac{-a_{12}a_{13} - a_{23}(a_{11} - 1)}{a_{12}^2 + (a_{11} - 1)^2}$$

giving a unique solution.

Case 2. $a_{11} = 1$. Then $a_{12} = 0$, since $a_{11}^2 + a_{12}^2 = +1$. So

$$AX = \begin{bmatrix} 1 & 0 & a_{13} \\ 0 & 1 & a_{23} \\ 0 & 0 & 1 \end{bmatrix} \begin{bmatrix} x_1 \\ x_2 \\ 1 \end{bmatrix} = \begin{bmatrix} x_1 + a_{13} \\ x_2 + a_{23} \\ 1 \end{bmatrix}$$

Thus there are no invariant points unless $a_{13} = a_{23} = 0$, in which case $A = I$. $\square$

Using the result of Theorem 3.12, direct isometries are defined to be either rotations or translations. (The identity transformation is defined as both.) As you may suspect, we shall show that these isometries do indeed have exactly the same properties as the rotations and translations described in other mathematics courses.

However, unlike the 2×2 matrix representations for rotations frequently used elsewhere, we will use 3×3 matrices. Furthermore, by using 3×3 matrices, we can also represent translations in matrix form.

Definition 3.15. The direct isometries with no invariant points together with the identity isometry are called *translations*. The direct isometries with exactly one invariant point together with the identity isometry are called *rotations*. The invariant point is called the *center* of the rotation.

According to this definiton, translations are those isometries whose matrix representation was determined in case 2 of the proof of Theorem 3.12. This matrix form is given explicity in the following theorem.

Theorem 3.13. *A translation T has matrix representation*

$$\begin{bmatrix} 1 & 0 & a \\ 0 & 1 & b \\ 0 & 0 & 1 \end{bmatrix}$$

T^{-1} *is also a translation and has matrix representation*

$$\begin{bmatrix} 1 & 0 & -a \\ 0 & 1 & -b \\ 0 & 0 & 1 \end{bmatrix}$$

The verification of the second half of the previous theorem as well as the next two theorems involve calculations with elementary matrix algebra (see Exercises 9–11).

Theorem 3.14. *The set of translations form a group.*

Theorem 3.15. *Given a point X and a point Y, there is a unique translation mapping X to Y.*

Using the matrix representation, several characteristic properties of translations can be identified. These properties should confirm the frequently used description of a translation as *sliding points along fixed lines.*

Theorem 3.16. *If a translation maps a line u to a line v, then u and v are either identical or parallel.*

Proof. If A is the matrix of the translation, then $kv = uA^{-1}$ since v is the image of u. So

$$k[v_1, v_2, v_3] = [u_1, u_2, u_3] \begin{bmatrix} 1 & 0 & -a \\ 0 & 1 & -b \\ 0 & 0 & 1 \end{bmatrix}$$

$$= [u_1, u_2, -au_1 - bu_2 + u_3].$$

Since $kv_1 = u_1$ and $kv_2 = u_2$ the conclusion follows (see Exercise 10 in Section 3.2). □

The proof of the following theorem demonstrates the interplay between synthetic and analytic methods and offers a nice application of many of the analytic techniques we have developed.

Theorem 3.17. *If a translation maps P to P′ (P ≠ P′), then the line PP′ as well as all lines parallel to PP′ are invariant. No other lines are invariant.*

Proof. Let P be a point with coordinates $(p_1, p_2, 1)$. If T is a translation with a matrix of the form given in Theorem 3.13, $P′ = T(P)$ has coordinates

$(p_1 + a, p_2 + b, 1)$ so the equation of line PP' is given by

$$\begin{vmatrix} x_1 & p_1 & p_1 + a \\ x_2 & p_2 & p_2 + b \\ 1 & 1 & 1 \end{vmatrix} = 0$$

Evaluating this determinant yields $[-b, a, p_1 b - p_2 a]$ as the line coordinates for PP'. So the lines parallel to PP', as well as PP', all have coordinates of the form $[-b, a, c]$. Applying the line equation of this translation gives

$$[-b, a, c] \begin{bmatrix} 1 & 0 & -a \\ 0 & 1 & -b \\ 0 & 0 & 1 \end{bmatrix} = [-b, a, c]$$

so these lines are indeed invariant under this translation.

To verify the second statement of the theorem, we will use an indirect proof. Assume a line l, not parallel to PP', is also invariant under the translation. Then l and PP' intersect at a point, Q. So Q is on two invariant lines, l and PP'. Since isometries preserve incidence $Q' = T(Q)$, must also be on both l and PP' implying that $Q = Q'$. This would mean that Q is an invariant point of the translation. However, since this translation is not the identity $(P \neq P')$, we have a contradiction. It follows that no lines in addition to those parallel to PP' are invariant. □

Theorem 3.18. *If u and v are parallel lines, then there is a translation mapping u to v.*

Proof. Let X be a point on u, X' a point on v. Then by Theorem 3.15 there is a translation maping X to X'. But this translation also maps line u to a line u' through X' (see Fig. 3.2). Since u' is parallel to u (Theorem 3.16), it follows that $u' = v$. □

Using the preceding properties of translations it is possible to determine synthetically the image of any other point under a translation T that maps P to P'. If Q is any other point, its image Q' can be found as follows.

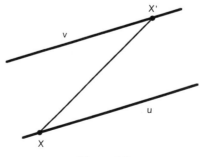

Figure 3.2

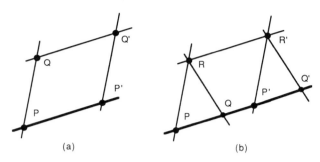

Figure 3.3

Case 1. Q is not on PP' (see Fig. 3.3a). Q' will be the point at which the line through P' parallel to PQ and the line through Q parallel to PP' intersect.

Case 2. Q is on PP' (see Fig. 3.3b). First find R' for some R not on PP' and then use R and R' in place of P and P' in case1.

Rotations too can be further characterized to allow a synthetic description of these mappings.

Theorem 3.19. *A rotation R with center $C(c_1, c_2, 1)$ has matrix representation:*

$$\begin{bmatrix} \cos \theta & -\sin \theta & c_1(1 - \cos \theta) + c_2 \sin \theta \\ \sin \theta & \cos \theta & -c_1 \sin \theta + c_2(1 - \cos \theta) \\ 0 & 0 & 1 \end{bmatrix}$$

and R^{-1} has matrix representation

$$\begin{bmatrix} \cos \theta & \sin \theta & c_1(1 - \cos \theta) - c_2 \sin \theta \\ -\sin \theta & \cos \theta & c_1(\sin \theta) + c_2(1 - \cos \theta) \\ 0 & 0 & 1 \end{bmatrix}$$

and is also a rotation with center C.

Proof. By definition, R has a matrix representation

$$\begin{bmatrix} a_{11} & a_{12} & a_{13} \\ -a_{12} & a_{11} & a_{23} \\ 0 & 0 & 1 \end{bmatrix} \quad \text{where } (a_{11})^2 + (a_{12})^2 = 1.$$

Since $|a_{11}| \leq 1$, we let $a_{11} = \cos \theta$, and it follows that $a_{12} = \pm \sin \theta$. Let $a_{12} = -\sin \theta$. We can find a_{13}, a_{23} by noting that $C(c_1, c_2, 1)$ must remain invariant.

So

$$\begin{bmatrix} \cos \theta & -\sin \theta & a_{13} \\ \sin \theta & \cos \theta & a_{23} \\ 0 & 0 & 1 \end{bmatrix} \begin{bmatrix} c_1 \\ c_2 \\ 1 \end{bmatrix} = \begin{bmatrix} c_1 \\ c_2 \\ 1 \end{bmatrix}$$

or

$$c_1 \cos \theta - c_2 \sin \theta + a_{13} = c_1$$
$$c_1 \sin \theta + c_2 \cos \theta + a_{23} = c_2$$

and solving for a_{13} and a_{23} gives the desired entries in the matrix for R. The matrix representation of R^{-1} can be found by inverting the matrix for R.

$$\square$$

Using these matrix representations, we can verify that for each point C, the rotations with center C form a group.

Theorem 3.20. *The set of all rotations with a given center C form a group.*

The θ in the matrix for R is called the *measure of the angle of rotation of R*, or more commonly, the *angle of rotation of R*. From the preceding theorem, it follows that a rotation is uniquely determined by its angle of rotation and its center so we can denote the rotation with center C and angle θ by $R_{C,\theta}$. Note also that $(R_{C,\theta})^{-1} = R_{C,-\theta}$; that is, the inverse of a rotation with angle θ is a rotation about the same center with angle $-\theta$. Although we can now refer to the angle of a rotation on the basis of its matrix representation, it is still necessary to verify that this angle characterizes the mapping defined by a rotation. This is done in Theorem 3.21. First, however, we make an observation that will be useful in the proof of this theorem, as well as in future calculations.

Even though Theorem 3.19 gives the matrix form for a rotation with any center C, it is sufficient to remember only the simpler form for rotations with center at the point $O(0,0,1)$. Then a translation T that maps O and C and a rotation with center O can be used as follows to find the rotation with center C (see Exercise 6):

$$R_{C,\theta} = T R_{O,\theta} T^{-1}$$

This equation makes it sufficient to verify the following theorem only for rotations with center $O(0,0,1)$ since the translations T and T^{-1} preserve angles.

Theorem 3.21. *Under a rotation with center C and angle θ, any point $P \neq C$ is mapped to a point P' such that $d(C,P) = d(C,P')$ and $m(\angle(PCP')) = \theta$.*

Proof. Since a rotation is an isometry and the images of points C and P under this isometry are C and P', respectively, it follows from the definition of isometries that $d(C,P) = d(C,P')$.

To verify that $m(\angle PCP') = \theta$, we will make use of the preceding observation and prove the result for a rotation with center $O(0,0,1)$. Assume P *has coordinates* $(p_1, p_2, 1)$. Then using the matrix representation of the rotation with center O and angle θ, we have

$$\begin{bmatrix} p'_1 \\ p'_2 \\ 1 \end{bmatrix} = \begin{bmatrix} \cos\theta & -\sin\theta & 0 \\ \sin\theta & \cos\theta & 0 \\ 0 & 0 & 1 \end{bmatrix} \begin{bmatrix} p_1 \\ p_2 \\ 1 \end{bmatrix}$$

Therefore P and P' have coordinates $(p_1, p_2, 1)$ and $(p_1\cos\theta - p_2\sin\theta, p_1\sin\theta + p_2\cos\theta, 1)$, respectively. It follows that the lines $u = OP$ and $v = OP'$ have the following coordinates: $u[-p_2, p_1, 0]$ and $v[-p_1\sin\theta - p_2\cos\theta, p_1\cos\theta - p_2\sin\theta, 0]$. Substituting these coordinates into the expression for $\tan(\angle(u,v))$ given in Definition 3.4, we get

$$\frac{u_1 v_2 - v_1 u_2}{u_1 v_1 + u_2 v_2} = \frac{-p_2(p_1\cos\theta - p_2\sin\theta) - (-p_1\sin\theta - p_2\cos\theta)p_1}{-p_2(-p_1\sin\theta - p_2\cos\theta) + p_1(p_1\cos\theta - p_2\sin\theta)}$$

$$= \frac{(p_1^2 + p_2^2)\sin\theta}{(p_1^2 + p_2^2)\cos\theta} = \tan\theta \qquad\qquad \square$$

Using this theorem the image of any point under a rotation $R_{C,\theta}$ can be determined synthetically, as shown in Fig. 3.4. This theorem also gives an immediate proof of the following theorem.

Theorem 3.22. *If lines u and v intersect at a point C with $m(\angle(u,v)) = \theta$ then $R_{C,\theta}$ will map u to v.*

Now that we have characterized the direct isometries, it is possible to prove the converse of Theorem 3.11 for triangles that have the same orientation.

Theorem 3.23. *If $\triangle PQR$ and $\triangle P'Q'R'$ are two triangles with $m(\overline{PQ}) = m(\overline{P'Q'})$, $m(\overline{QR}) = m(\overline{Q'R'})$, $m(\overline{RP}) = m(\overline{R'P'})$, $m(\angle PQR) = m(\angle P'Q'R')$, $m(\angle QRP) = m(\angle Q'R'P')$, and $m(\angle RPQ) = m(\angle R'P'Q')$, then there is a direct isometry mapping $\triangle PQR$ to $\triangle P'Q'R'$ so $\triangle PQR \simeq \triangle P'Q'R'$.*

Proof. To show the congruence of the two triangles, it is sufficient to show that there is an isometry mapping $\triangle PQR$ to $\triangle P'Q'R'$ (see Definition 3.12). In the following paragraph we will outline a procedure for obtaining such an isometry.

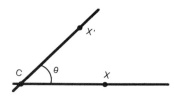

Figure 3.4

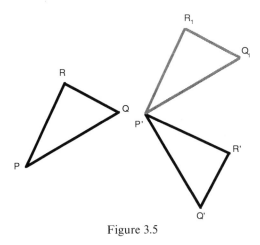

Figure 3.5

Let T be a translation mapping P to P'. T will also map points Q and R to points Q_1 and R_1 as indicated in Fig. 3.5. Let $\theta = m(\angle Q_1 P'Q')$. Then the rotation with center P' and angle θ will map the points P' and Q_1 to P' and Q', respectively, since $d(P', Q_1) = d(P, Q) = d(P', Q')$. Furthermore since $m(\angle Q_1 P'R_1) = m(\angle QPR) = m(\angle Q'P'R')$ and $d(P', R_1) = d(P, R) = d(P', R')$, this rotation will also map point R_1 to R'. Therefore the isometry consisting of the composite $R_{P',\theta} T$ will map $\triangle PQR$ to $\triangle P'Q'R'$.   □

EXERCISES

1. Let T be the translation mapping $X(1, -2, 1)$ to $X'(3, 4, 1)$. (a) Find the matrix of T and the image of line $u[2, 3, -1]$ under T. (b) Verify that lines u and $T(u)$ are parallel.

2. Find the matrix of a translation which maps $u[1, -2, 5]$ to $v[2, -4, 7]$.

3. Find the invariant lines of the translation with matrix

$$\begin{bmatrix} 1 & 0 & -3 \\ 0 & 1 & 7 \\ 0 & 0 & 1 \end{bmatrix}$$

4. Prove: If l is a line, the set of all translations that keep l invariant forms a group.

5. (a) Verify that the following matrix is a matrix of a rotation:

$$\begin{bmatrix} \frac{3}{5} & \frac{4}{5} & 2 \\ -\frac{4}{5} & \frac{3}{5} & 1 \\ 0 & 0 & 1 \end{bmatrix}$$

(b) What is the angle of this rotation? (c) What is the center of this rotation?

6. Verify that $TR_{0,\theta}T^{-1}$ is a rotation with center C and angle θ where O is the point with coordinates $(0, 0, 1)$ and T is a translation mapping O to C.

7. (a) Find the point of intersection of lines $u[2, 0, 3]$ and $v[1, 1, 5]$. (b) Find $m(\angle(u, v))$. (c) Find a rotation that maps line u to line v. (Be sure to check your answer.)

8. (a) Describe synthetically how to find the center and angle of a rotation that maps a given point P to a given point P'. (Note that there are an infinite number of possible rotations.) (b) Use the answer to part (a) to find the matrix of a rotation that maps $P(2, 0, 1)$ to $P'(1, -3, 1)$. Verify that the rotation works.

9. Prove that T^{-1} has a matrix representation of the form given in Theorem 3.13.

10. Prove Theorem 3.14.

11. Prove Theorem 3.15.

12. Give an analytic proof of the second statement in Theorem 3.17 (no other lines are invariant).

13. (a) Demonstrate by diagramming an example that the product of two rotations with different centers can be a translation. (b) Show how one particular triangle is mapped under the two rotations you used in part (a).

14. Show that $R_{C,\theta}R_{C,\phi} = R_{C,\phi+\theta}$ by using matrices.

15. Prove that any direct isometry can be expressed as the product of a rotation with center $O(0, 0, 1)$ and a translation (the wording indicates that the rotation is to be used first and the translation second).

16. (a) Find a direct isometry that maps $P(1, 0, 1)$ and $Q(5, 3, 1)$ to $P'(3, -2, 1)$ and $Q'(0, 2, 1)$, respectively. (b) Is the isometry you found a rotation or a translation? Why?

17. (a) Use the procedure outlined in the proof of Theorem 3.23 to verify that $\triangle PQR$ and $\triangle P'Q'R'$ are congruent where P, Q, R, P', Q', and R' have the following coordinates: $P(2, 8, 1)$, $Q(4, 4, 1)$, $R(10, 7, 1)$, $P'(7, -2, 1)$, $Q'(11, -4, 1)$, and $R'(14, 2, 1)$. (b) Verify that the isometry you found in part (a) is a rotation. What is its center?

18. Prove: If $C \neq D$ and $R_{C,\theta}$ and $R_{D,\phi}$ are two nonidentity rotations, then the product $R_{D,\phi}R_{C,\theta}$ is (a) a rotation with angle $\theta + \phi$ if $\theta + \phi \neq 0 \pmod{2\pi}$ and (b) a translation if $\theta + \phi = 0 \pmod{2\pi}$.

19. Prove: If T is a direct isometry such that $T^2 = I$, where I is the identity, then $T = I$ or T is a rotation with an angle of $180°$.

20. Prove: (a) If a line is invariant under a nonidentity rotation with center C, then the line is incident with C. [Hint: Assume $C = O(0, 0, 1)$.] (b) If a

nonidentity rotation with center C has an invariant line, then the angle of the rotation is $180°$.

3.6. Indirect Isometries

In this section we will find that, as in the case of direct isometries, there are two types of indirect isometries: those that have invariant points and those that do not. In this case, however, there are indirect isometries that keep not just one point but *every* point on a particular line invariant. Such a line is said to be *pointwise invariant*. The adjective "pointwise" is important, since a line can be invariant without any points on it being invariant. An example of this is the line PQ $(P \neq Q)$ under a translation that maps P to Q. Line PQ is invariant under the translation by Theorem 3.17, but since translations have no invariant points, none of the points on PQ remain invariant.

Definition 3.16. A *reflection with axis m*, denoted R_m, is an indirect isometry that keeps line m pointwise invariant.

Having defined reflections, it is important to verify that such transformations exist. In particular, we find the matrix of the reflection R_x where x is the line with coordinates $[0, 1, 0]$, that is, the line known as the x- or x_1-axis. Then, using this matrix, the matrix for a reflection in any other axis m can be obtained.

Theorem 3.24. *The matrix representation of a reflection R_x with axis $x[0, 1, 0]$ is*

$$\begin{bmatrix} 1 & 0 & 0 \\ 0 & -1 & 0 \\ 0 & 0 & 1 \end{bmatrix}$$

In general the matrix representation of a reflection R_m can be found using $R_m = SR_xS^{-1}$ where S is a direct isometry mapping x to m $(S(x) = m)$.

Proof. All points on x have coordinates of the form $(x_1, 0, 1)$. R_x is then an indirect isometry that keeps each point $(x_1, 0, 1)$ fixed, that is, for all $x_1 \in R$

$$\begin{bmatrix} a_{11} & a_{12} & a_{13} \\ a_{12} & -a_{11} & a_{23} \\ 0 & 0 & 1 \end{bmatrix} \begin{bmatrix} x_1 \\ 0 \\ 1 \end{bmatrix} = \begin{bmatrix} x_1 \\ 0 \\ 1 \end{bmatrix}$$

So $a_{11}x_1 + a_{13} = x_1$ and $a_{12}x_1 + a_{23} = 0$. Since this must hold for all $x_1 \in R$, it follows that $a_{11} = 1$ and $a_{13} = a_{12} = a_{23} = 0$. So the matrix for R_x is of the form given.

In general, there is always a direct isometry S, mapping x to m (see Theorems 3.18 and 3.22). By the corollary to Theorem 3.8, SR_xS^{-1} is an indirect isometry. Then it is sufficient to show that SR_xS^{-1} keeps m pointwise invariant. If X is an arbitrary point on m, $S^{-1}(X)$ is a point on line x, so $R_x(S^{-1}(X)) = S^{-1}(X)$. Thus $(SR_xS^{-1})(X) = S(S^{-1}(X)) = X$ and so X remains invariant. □

Corollary. $R_m^{-1} = R_m$.

According to the definition, a reflection with axis m keeps all points on m invariant. But does R_m have any other invariant points? The answer to this question is given by the next theorem.

Theorem 3.25. *The only invariant points under a reflection with axis m, R_m, are those on m. For any point P not on m, if $P' = R_m(P)$, then $P' \neq P$ and m is the perpendicular bisector of segment PP'.*

Proof. Let P be a point with coordinates $(p_1, p_2, 1)$. Then $P' = R_x(P)$ has coordinates $(p_1, -p_2, 1)$ so the invariant points of R_x are precisely those for which $p_2 = -p_2$ or $p_2 = 0$. These are the points on $x[0, 1, 0]$. Using this result for the reflection R_x, it is then possible to verify that the only invariant points of $R_m = SR_xS^{-1}$ are those on line m (see Exercise 1).

To show that m is the perpendicular bisector of the line PP' where P is not on m, we will again first verify the result for the case where m is the line x (see Fig. 3.6). Since $P(p_1, p_2, 1)$ is not on m, $p_2 \neq 0$. The line $p = PP'$ has coordinates $[1, 0, -p_1]$ and x has coordinates $[0, 1, 0]$, so using Definition 3.4, we get $m(\angle(p, x)) = 90°$. Also, lines p and x intersect at the point $Q(p_1, 0, 1)$ so $d(P, Q) = d(P', Q) = |p_2|$ and it follows that x bisects $\overline{PP'}$. In general, the same results can be shown for any line m by writing $R_m = SR_xS^{-1}$ (where S is a direct isometry mapping x to m) and noting that S preserves angle measure and distance. □

The previous theorem suggests that lines perpendicular to a line m are invariant under a reflection with axis m. The proof of this and the remaining theorems about reflections will be given for the case where the axis of the reflection is $x[0, 1, 0]$. As outlined in the proof of Theorem 3.25, these results

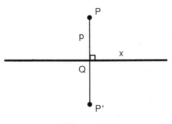

Figure 3.6

can then be extended to reflections with other axes by writing $R_m = SR_xS^{-1}$ where S is a direct isometry mapping x to m.

Theorem 3.26. *Every line perpendicular to m is invariant under R_m, and conversely any line invariant under R_m is either m or a line perpendicular to m.*

Proof. As noted earlier, we will assume that $m = x$. Let u be a line perpendicular to x. Since x has coordinates $[0, 1, 0]$ it follows from Definition 3.4 that u has coordinates $[u_1, 0, u_3]$. If u' is the image of u under the reflection, then $ku' = u(R_x)^{-1} = uR_x$ or, in matrix notation,

$$ku' = [u_1, 0, u_3] \begin{bmatrix} 1 & 0 & 0 \\ 0 & -1 & 0 \\ 0 & 0 & 1 \end{bmatrix} = [u_1, 0, u_3]$$

So $u' = u$ is invariant.

Now assume $u[u_1, u_2, u_3]$ is an invariant line under R_x. Then

$$k[u_1, u_2, u_3] = [u_1, u_2, u_3] \begin{bmatrix} 1 & 0 & 0 \\ 0 & -1 & 0 \\ 0 & 0 & 1 \end{bmatrix}$$

So $ku_1 = u_1$, $ku_2 = -u_2$, and $ku_3 = u_3$. If u_1 or $u_3 \neq 0$, it follows that $k = 1$ and hence $u_2 = 0$. Thus u has coordinates $[u_1, 0, u_3]$ and so is perpendicular to x. If $u_1 = u_3 = 0$, $u_2 \neq 0$, and so $u = x[0, 1, 0]$.   □

Using reflections along with direct isometries it is now possible to verify the converse of Theorem 3.11 for triangles with opposite orientation (see Exercise 4).

Theorem 3.27. *If $\triangle PQR$ and $\triangle P'Q'R'$ are two triangles with $m(\overline{PQ}) = m(\overline{P'Q'})$, $m(\overline{QR}) = m(\overline{Q'R'})$, $m(\overline{RP}) = m(\overline{R'P'})$, and also $m(\angle PQR) = -m(\angle P'Q'R')$, $m(\angle QRP) = -m(\angle Q'R'P')$, and $m(\angle RPQ) = -m(\angle R'P'Q')$, then there is an indirect isometry mapping $\triangle PQR$ to $\triangle P'Q'R'$ so $\triangle PQR \simeq \triangle P'Q'R'$.*

The properties of reflections described previously also lead to the remarkable fact that the direct isometries used in the proof of the preceding theorem are themselves products of two reflections with appropriately chosen axes.

Theorem 3.28. *The product R_nR_m of two reflections with axes m and n is (a) a translation mapping any point P to a point P' where $d^*(P, P') = 2d^*(m, n)$ if n and m are parallel (d^* indicates directed distance); or (b) a rotation with center C and angle $\theta = 2[m(\angle(m, n))]$ if n intersects m at point C. Conversely, any translation or rotation can be written as the product R_nR_m where lines m and n have the properties described in (a) or (b), respectively.*

Proof. (a) We shall assume that m is the line with coordinates $[0, 1, 0]$; n must then have coordinates $[0, 1, n_2]$. We can measure the (perpendicular) distance between m and n along the line $t[1, 0, 0]$, which is perpendicular to m and n at $M(0, 0, 1)$ and $N(0, -n_2, 1)$, respectively. Then $d*(m, n) = d*(M, N) = -n_2$. To find the matrix representation of R_n, we will use the translation T, which maps M to N and hence line m to line n. The matrix of T is

$$T = \begin{bmatrix} 1 & 0 & 0 \\ 0 & 1 & -n_2 \\ 0 & 0 & 1 \end{bmatrix}$$

So $R_n = TR_mT^{-1}$ has matrix representation

$$\begin{bmatrix} 1 & 0 & 0 \\ 0 & 1 & -n_2 \\ 0 & 0 & 1 \end{bmatrix}\begin{bmatrix} 1 & 0 & 0 \\ 0 & -1 & 0 \\ 0 & 0 & 1 \end{bmatrix}\begin{bmatrix} 1 & 0 & 0 \\ 0 & 1 & n_2 \\ 0 & 0 & 1 \end{bmatrix} = \begin{bmatrix} 1 & 0 & 0 \\ 0 & -1 & -2n_2 \\ 0 & 0 & 1 \end{bmatrix}$$

Thus $R_n R_m$ has matrix representation

$$\begin{bmatrix} 1 & 0 & 0 \\ 0 & -1 & -2n_2 \\ 0 & 0 & 1 \end{bmatrix}\begin{bmatrix} 1 & 0 & 0 \\ 0 & -1 & 0 \\ 0 & 0 & 1 \end{bmatrix} = \begin{bmatrix} 1 & 0 & 0 \\ 0 & 1 & -2n_2 \\ 0 & 0 & 1 \end{bmatrix}$$

which is the matrix of a translation. Furthermore

$$X' = \begin{bmatrix} 1 & 0 & 0 \\ 0 & 1 & -2n_2 \\ 0 & 0 & 1 \end{bmatrix}\begin{bmatrix} x_1 \\ x_2 \\ 1 \end{bmatrix} = \begin{bmatrix} x_1 \\ x_2 - 2n_2 \\ 1 \end{bmatrix}$$

Thus $d*(X, X') = -2n_2 = d*(m, n)$.

(b) The proof is similar to that for (a) except in this case it is necessary to use a rotation to map m to n (see Exercise 6).

Since the proof of the second half of the theorem also involves two cases, we will again prove one case and leave the other case as an exercise (see Exercise 7).

To verify the converse of (b), let $R_{C, 2\theta}$ be a rotation with center C. To simplify the matrix computations we will assume $C = O(0, 0, 1)$. Then $R_{O, 2\theta}$ has matrix representation

$$\begin{bmatrix} \cos(2\theta) & -\sin(2\theta) & 0 \\ \sin(2\theta) & \cos(2\theta) & 0 \\ 0 & 0 & 1 \end{bmatrix}$$

Now consider the reflections R_m and R_n where $m[0, 1, 0]$ and $n[-\sin\theta, \cos\theta, 0]$. To find the matrix representation for R_n, note that m and n intersect at point O. Then

$$R_{O, \theta} = \begin{bmatrix} \cos\theta & -\sin\theta & 0 \\ \sin\theta & \cos\theta & 0 \\ 0 & 0 & 1 \end{bmatrix}$$

is a rotation with center O which maps m to n, and R_n has matrix representation

$$\begin{bmatrix} \cos\theta & -\sin\theta & 0 \\ \sin\theta & \cos\theta & 0 \\ 0 & 0 & 1 \end{bmatrix} \begin{bmatrix} 1 & 0 & 0 \\ 0 & -1 & 0 \\ 0 & 0 & 1 \end{bmatrix} \begin{bmatrix} \cos\theta & \sin\theta & 0 \\ -\sin\theta & \cos\theta & 0 \\ 0 & 0 & 1 \end{bmatrix}$$

$$= \begin{bmatrix} \cos^2\theta - \sin^2\theta & 2(\sin\theta)(\cos\theta) & 0 \\ 2(\sin\theta)(\cos\theta) & \sin^2\theta - \cos^2\theta & 0 \\ 0 & 0 & 1 \end{bmatrix}$$

$$= \begin{bmatrix} \cos(2\theta) & \sin(2\theta) & 0 \\ \sin(2\theta) & -\cos(2\theta) & 0 \\ 0 & 0 & 1 \end{bmatrix}$$

Finally $R_n R_m$ has matrix representation

$$\begin{bmatrix} \cos(2\theta) & -\sin(2\theta) & 0 \\ \sin(2\theta) & \cos(2\theta) & 0 \\ 0 & 0 & 1 \end{bmatrix} = R_{O,2\theta}$$ □

Corollary 1. *A direct isometry is the product of two reflections.*

Corollary 2. $R_m R_n = R_{m'} R_{n'}$ *iff* m, n, m', n' *are all parallel and* $d^*(m,n) = d^*(m',n')$ *or* m, n, m', n' *are all concurrent and* $m(\angle(m,n)) = m(\angle(m',n'))$.

Corollary 1 not only summarizes Theorem 3.28 but should also suggest the following question: Are the indirect isometries also products of reflections? Clearly, reflections themselves are indirect isometries. There are, however, other indirect isometries and the remainder of this section is devoted to the study of these "other" indirect isometries.

Theorem 3.29. *An indirect isometry is the product of one or three reflections.*

Proof. The proof follows easily after noting that any indirect isometry can be expressed as a product of a direct isometry and R_x (see Exercise 9).   □

The next theorem demonstrates that the indirect isometries are either reflections or glide reflections. The proof of this result makes extensive use of Corollary 2 to Theorem 3.28.

Figure 3.7

Definition 3.17. A *glide reflection with axis m* (see Fig. 3.7) is the product of a reflection with axis m and a nonidentity translation along m (i.e., m is invariant under the translation).

Theorem 3.30. *An indirect isometry is either a reflection or a glide reflection.*

Proof. By Theorem 3.29 we need only consider indirect isometries that can be written as the product $R_c R_b R_a$. We will examine this product for each of several cases.

Case 1. a, b, c are all parallel (see Fig. 3.8). Then let c' be a line parallel to a such that $d^*(c', a) = d^*(c, b)$. By Corollary 2 $R_c R_b = R_{c'} R_a$, so $R_c R_b R_a = R_{c'} R_a R_a = R_{c'}$.

Case 2. a, b, c are concurrent (see Fig. 3.9). The proof is similar to that in case 1 except here the line c' must be chosen so that c' is a line through $a \cdot b$ and $m(\angle(c', a)) = m(\angle(c, b))$.

Case 3. Each pair of the lines a, b, c intersects in distinct ordinary points (see Fig. 3.10). Let b' be incident with $b \cdot c$ and perpendicular to a. Let c' be a line incident with $b \cdot c$ (the point of intersection of lines b and c) such that $m(\angle(c, b)) = m(\angle(c', b'))$. Then $R_c R_b R_a = R_{c'} R_{b'} R_a$ where b' is perpendicular to a. Let a' be a line incident with $a \cdot b'$ and perpendicular to c' and let b'' be a line incident with $a \cdot b'$ such that $m(\angle(a, b')) = m(\angle(a', b''))$. Then $R_{c'} R_{b'} R_a = R_{c'} R_{b''} R_{a'}$ where a' is perpendicular to c'. Since b' is perpendicular to a, it follows that b'' is perpendicular to a', so c' and b'' are parallel and both perpendicular to a'. Therefore $R_{c'} R_{b''}$ is a translation along a', so that the product $R_{c'} R_{b''} R_{a'}$ is a glide reflection with axis a'.

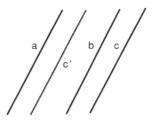

Figure 3.8

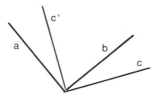

Figure 3.9

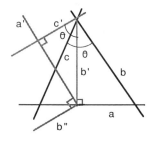

Figure 3.10

Case 4. Exactly two of the lines are parallel. The proof for this case is similar to the preceding ones.   □

We have now shown that there are indeed two types of indirect isometries. That reflections keep a line pointwise invariant has already been noted. Glide reflections, on the other hand, have an invariant line out no invariant points (see Exercise 16).

Theorem 3.31. *A glide reflection has one invariant line but no invariant points.*

With the results of this section we can now completely categorize the isometries and show that the isometry mapping a triangle to a congruent triangle is unique (see Exercises 17–19).

Theorem 3.32. *Every isometry is either a rotation, a translation, a reflection, or a glide reflection; and therefore every isometry can be written as the product of at most three reflections.*

Corollary. *An isometry with three noncollinear invariant points is the identity.*

Theorem 3.33. *If △ PQR ≃ △ P'Q'R', then there is a unique isometry mapping △ PQR onto △ P'Q'R' where P maps to P', Q to Q', and R to R'.*

Exercises

1. If S is a direct isometry mapping line x to line m, show that the *only* invariant points of $R_m = S R_x S^{-1}$ are the points on m.

2. (a) Find the matrix of R_m where m is the line $x_2 = (\sqrt{3}/3)x_1$. (b) Use this matrix to find P', the image of the point $P(3, 7, 1)$ under this reflection. (c) Verify that m is the perpendicular bisector of $\overline{PP'}$.

3. The proof of Theorem 3.26 uses an analytic argument to show that lines perpendicular to m are invariant under a reflection with axis m. Give a synthetic proof of the same result.

4. Outline a proof of Theorem 3.27 similar to the outline given for the proof of Theorem 3.23 in Section 3.5.

5. Find a product of a translation, a rotation, and a reflection that maps $\triangle PQR$ to $\triangle P'Q'R'$ where $P(-2, 5, 1)$, $Q(-2, 7, 1)$, $R(-5, 5, 1)$, $P'(4, 3, 1)$, $Q'(6, 3, 1)$, and $R'(4, 0, 1)$.

6. Prove part b of Theorem 3.28.

7. Prove the converse of part a of Theorem 3.28.

8. Let R_a and R_b be reflections with axes a and b where $a \neq b$. Prove: $R_a R_b = R_b R_a$ if and only if a and b are perpendicular. (*Hint:* Use Theorem 3.28.]

9. Prove Theorem 3.29.

10. Show by diagramming an example, that the product of two glide reflections can be a translation. (Be sure to indicate how a triangle is mapped under the glide reflections.)

11. Identify the following matrices as matrices of reflections or glide reflections. In each case find the axis.

$$\text{(a)} \quad \begin{bmatrix} 0 & 1 & -1 \\ 1 & 0 & 1 \\ 0 & 0 & 1 \end{bmatrix} \qquad \text{(b)} \quad \begin{bmatrix} 0 & 1 & 3 \\ 1 & 0 & 1 \\ 0 & 0 & 1 \end{bmatrix}$$

$$\text{(c)} \quad \frac{1}{25} \begin{bmatrix} -7 & -24 & -64 \\ -24 & 7 & -48 \\ 0 & 0 & 25 \end{bmatrix}$$

12. Show that if l and m are two parallel lines and n is the image of line l under a glide reflection with axis m, then l and n are also parallel.

13. Let P' be the image of P under a glide reflection with axis m. Show that m bisects the segment $\overline{PP'}$.

14. (a) Find the matrix of a glide reflection which maps $\overline{PQ}$ to $\overline{P'Q'}$ where $P(4, -2, 1)$, $Q(7, 2, 1)$, $P'(-3, -4, 1)$, and $Q'(0, 0, 1)$. (*Hint:* Use the result of Exercise 13.] (b) Is this glide reflection unique? Why? (c) Is there a unique glide reflection mapping P to P'? Why?

15. Prove: If line a intersects parallel lines b and c, then $R_c R_b R_a$ is a glide reflection. (This is part of case 4 in the proof of Theorem 3.30.)

16. Prove Theorem 3.31.

17. Prove Theorem 3.32.

18. Prove the corollary to Theorem 3.32.

19. Prove Theorem 3.33. [*Hint*: Assume there are two different isometries and use the corollary to Theorem 3.32 to obtain a contradiction.]

3.7. Symmetry Groups

The study of the invariants of isometries allows us to determine the structure of the Euclidean plane, since isometries are correspondences of the whole Euclidean plane with itself. Likewise, by considering the invariants of isometries that leave a figure unchanged, we can determine much about the structure of that figure. A *symmetry* transformation is an isometry that is identified with a specific figure. For example, the rotation $R = R_{C,90}$ maps each point of the design shown in Fig. 3.11 onto another point of the design. Furthermore, each rotation in the set $\{I, R, R^2, R^3\}$ is also a symmetry of this figure. This group of symmetries is known as the *symmetry group* of this figure.

Definition 3.18. If α is a set of points and T is an isometry such that $T(\alpha) = \alpha$, then T is a *symmetry* of α.

Theorem 3.34. *The set of all symmetries of a set of points forms a group.*

To find the symmetry group of a line segment $\overline{PQ}$, we must determine which isometries keep $\overline{PQ}$ invariant. Other than the identity, the only direct isometry that has this property is the rotation with center at the midpoint of $\overline{PQ}$ and angle 180°. Such rotations are known as *half-turns*.

Definition 3.19. A rotation with center C and angle 180° is a *half-turn with center C*, denoted H_C.

The only indirect symmetries of $\overline{PQ}$ are the two reflections with axes $m = PQ$ and n where n is the perpendicular bisector of $\overline{PQ}$. Other figures may have more symmetries, but half-turns and reflections play an important role in any discussion of symmetry as indicated by the following definition.

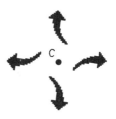

Figure 3.11

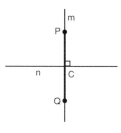

Figure 3.12

Definition 3.20. Let α be a set of points. If H_P is a half-turn with center P such that $H_P(\alpha) = \alpha$, then P is a *point of symmetry* for α. If R_m is a reflection with axis m such that $R_m(\alpha) = \alpha$, then m is a *line of symmetry* of α.

Thus a segment $\overline{PQ}$ has a point of symmetry, namely, its midpoint C, and two lines of symmetry; the lines $m = PQ$ and n, the perpendicular bisector of $\overline{PQ}$ (see Fig. 3.12).

The set of all symmetries of $\overline{PQ}$ is therefore $S = \{I, R_m, R_n, H_C\}$. The set S is closed under composition since the product of any two elements in S is again in S. This is demonstrated by Table 3.1, where an entry in a cell in the row labeled R_m and column labeled H_C indicates the product $R_m H_C$. This table also demonstrates that the inverse of each element in S is also an element of S, so S is a group.

Other geometric figures also have corresponding symmetry groups. In particular, every regular polygon has a finite symmetry group.

Definition 3.21. If a group G has exactly n elements, then G is *finite* and n is the *order* of G. If G is not finite, it is *infinite*.

To determine the finite symmetry group for an equilateral triangle, we note that such a triangle has three lines of symmetry as shown in Fig. 3.13, but no point of symmetry.

If m, n, and o are the lines of symmetry as shown in Fig. 3.13, then the identity transformation, as well as the reflections R_m, R_n, and R_o, are symmetries of the equilateral triangle. In addition, the rotations through

Table 3.1. Symmetries of a Segment

	I	R_m	R_n	H_C
I	I	R_m	R_n	H_C
R_m	R_m	I	H_C	R_n
R_n	R_n	H_C	I	R_m
H_C	H_C	R_n	R_m	I

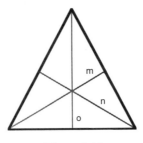

Figure 3.13

angles of $120°$ and $240°$ obtained as products of any two of these reflections are symmetries of the triangle. Using Theorem 3.28, we can find the products of each pair of these symmetries, as displayed in Table 3.2.

Using this multiplication table it is possible to show that each of the symmetries of an equilateral triangle can be obtained using products of R_m and R_{120}; that is, the symmetry group is generated by these two symmetries (see Exercise 7).

Definition 3.22. If every element of a group G is a product of the elements $T_1, T_2, \ldots, T_n$ then G is *generated* by $T_1, T_2, \ldots, T_n$, denoted by $G = \langle T_1, T_2, \ldots, T_n \rangle$.

A similar procedure can be used to determine the finite symmetry group for any regular polygon. Since these symmetry groups cannot contain a nonidentity translation or glide reflection, their only elements are rotations and reflections. It is possible to show that any of these finite groups can be generated by either one or two symmetries. For this reason, they are known as *cyclic* and *dihedral* groups, respectively (see Exercises 9 and 10).

There are figures whose symmetry groups do include translations. A pattern characterized by its remaining invariant under some shortest nonidentity translation is known as a *frieze* pattern. The term "frieze pattern" refers to the occurrence of these patterns as a repeated motif around the frieze of older

Table 3.2. Symmetries of an Equilateral Triangle

	I	R_m	R_n	R_0	R_{120}	R_{240}
I	I	R_m	R_n	R_0	R_{120}	R_{240}
R_m	R_m	I	R_{120}	R_{240}	R_n	R_0
R_n	R_n	R_{240}	I	R_{120}	R_0	R_m
R_0	R_0	R_{120}	R_{240}	I	R_m	R_n
R_{120}	R_{120}	R_0	R_m	R_n	R_{240}	I
R_{240}	R_{240}	R_n	R_0	R_m	I	R_{120}

Figure 3.14

buildings. An example of a frieze pattern as reprinted from Audsley's (1968) *Designs and Patterns from Historic Ornament* is shown in Fig. 3.14. Frieze patterns clearly contain lines that are invariant under their defining translation.

Definition 3.23. A group of isometries that keeps a given line c invariant and whose translations form an infinite cyclic subgroup is a *frieze group with center c*.

The translation that generates the cyclic subgroup of a frieze group is the "shortest" translation referred to in the description of frieze patterns (see Exercise 12). In the remainder of this section, we will denote this shortest translation by τ.

Definition 3.24. If $T_{A,B}$ is a translation mapping A to B, then $d(A, B)$ is called the *length* of the translation. If $d(A, B) < d(C, D)$, then $T_{A,B}$ is *shorter* than $T_{C,D}$.

To find the possible frieze groups with center c that contain a translation τ, we need to decide which isometries are candidates for elements of such a group. In addition to the translations in $\langle \tau \rangle$ the only permissible nonidentity direct isometries are the half-turns with centers on c. The permissible indirect isometries are the reflection in c, the reflections in lines perpendicular to c, and glide reflections along c. Using all possible combinations of these as generators along with τ, we can compile an exhaustive list of the frieze groups G_i with center c. By determining which half-turns and reflections belong to each frieze group G_i, we can find the points and lines of symmetry of the frieze pattern having G_i as its symmetry group. Unlike polygons, frieze patterns may have an infinite number of such points and lines (see Exercise 11).

The procedure we use in making this list makes use of the following notation. If G_i contains half-turns, we let P denote the center of a half-turn. If G_i contains no half-turns but does contain reflections in lines perpendicular to c, p denotes the axis of one such reflection and P is the point of intersection of the lines p and c. Otherwise we choose P to be any point on c. We let $P_n = \tau^n(P)$ where τ^n denotes the composition of τ with itself n times for nonnegative integers n and the composition of τ^{-1} with itself n times for negative integers n. $M = M_o$ denotes the midpoint of $\overline{P_o P_1}$ with $M_n = \tau^n(M)$. Note that M_n is also the midpoint of $\overline{P_n P_{n+1}}$ (see Fig. 3.15). In order to streamline the presentation, the justification of many of the steps is relegated to Exercises 14–18.

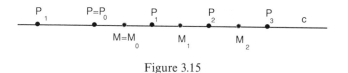

Figure 3.15

The Seven Possible Frieze Groups

Frieze Groups Containing Only Direct Isometries

1. $G_1 = \langle \tau \rangle$: The group generated by only τ, the shortest translation, contains no half-turns, reflections, or glide reflections, so the associated frieze pattern, F_1, has no points or lines of symmetry. Furthermore, there is no glide reflection which keeps F_1 invariant.
2. $G_2 = \langle \tau, H_P \rangle$: The group generated by τ and a half-turn about a point P on c also contains all products $\tau^n H_P$. Each of these is again a half-turn with $\tau^{2k} H_P$ and $\tau^{2k+1} H_P$ yielding the half-turns with centers at P_k and M_k, respectively.

To show that G_2 is the only frieze group of direct isometries containing half-turns, it is sufficient to show that any half-turns other than H_P must be one of those in G_2. If H_C is an arbitrary half-turn with center $C \neq P$ on line c, then the translation $H_C H_P$ is also in G_2. However, all translations in G_2 are in the cyclic subgroup $\langle \tau \rangle$. Thus, for some integer n, $H_C H_P(P) = P_n$, or $P = H_P(P) = H_C(P_n)$. This implies that C is the midpoint of $\overline{PP_n}$ so C is either P_k or M_k for some integer k.

Since G_2 contains the half-turn H_P but no indirect isometries, the associated frieze pattern, F_2, has a point of symmetry but no line of symmetry.

Frieze Groups Containing Indirect Isometries. We first consider frieze groups that contain reflections but do *not* contain half-turns. This means that these groups cannot contain both the reflections R_c and R_p where p is a line perpendicular to c at P, since $R_c R_p = H_P$.

3. $G_3 = \langle \tau, R_c \rangle$: Since the reflection with axis c and the translation τ have the properties that $R_c R_c = I$ and $\tau R_c = R_c \tau$, the other elements in the group generated by τ and R_c will consist of products of the form $\tau^n R_c$. These are all glide reflections with axis c mapping P to P_n. Thus, the associated frieze pattern, F_3, has its center c as a line of symmetry but no point of symmetry.
4. $G_4 = \langle \tau, R_p \rangle$: Since the reflection with axis p and the translation τ have the properties that $R_p R_p = I$ and $R_p \tau = \tau^{-1} R_p$, the other elements of this group will consist of products of the form $\tau^n R_p$. These are all reflections with $\tau^{2k} R_p$ and $\tau^{2k+1} R_p$ producing reflections in the lines perpendicular to c at P_k and M_k, respectively.

We can show that G_4 is the only frieze group containing reflections with axis perpendicular to c by demonstrating that any reflection with an axis perpendicular to c must be one of those in G_4. Let R_q be a reflection with axis $q \neq p$ perpendicular to c at Q. Then the product $R_q R_p$ is a translation along c and thus for some n, $R_q R_p(P) = P_n$ or $P = R_p(P) = R_q(P_n)$ so Q is the midpoint of $\overline{PP_n}$. This implies that Q is either P_k of M_k for some integer k.

Thus the associated frieze pattern, F_4, has a line perpendicular to its center as a line of symmetry but no point of symmetry.

If we allow half-turns and reflections, we obtain two more frieze groups. The final frieze group contains glide reflections, but no reflections. These three groups are described briefly here. A more comprehensive description is given in Martin (1982b, Chap. 10).

5. $G_5 = \langle \tau, H_P, R_c \rangle$: Since $H_P R_c = R_p R_c R_c = R_p$ the reflection R_p is in G_5. G_5 also includes elements of the form $\tau^n R_c$, which are glide reflections mapping P to P_n. Other elements are $\tau^{2k} R_p$ and $\tau^{2k+1} R_p$, reflections in lines perpendicular to c at P_k and M_k, respectively. The associated frieze pattern, F_5, has a point of symmetry. It also has the center line and a line perpendicular to the center as lines of symmetry.

6. $G_6 = \langle \tau, H_P, R_q \rangle$: To obtain a frieze group different from G_5, R_q must be a reflection in a line perpendicular to c at a point Q distinct from the points P_k and M_k. As Martin shows, the point Q must be the midpoint of $\overline{PM_k}$ for some k. The associated frieze pattern has a point of symmetry and a line of symmetry where the line of symmetry is perpendicular to the center.

7. $G_7 = \langle \sigma \rangle$: G_7 is generated by the glide reflection σ where $\sigma^2 = \tau$. Since G_7 contains no reflections or half-turns, the associated frieze pattern, F_7, has no point of symmetry and no line of symmetry. Unlike G_1 it does remain invariant under a glide reflection.

The seven possible frieze patterns determined by these frieze groups along with their types of symmetry are displayed in Table 3.3.

Just as frieze patterns remain invariant under a group generated by one

Table 3.3

Frieze Pattern		Type of Symmetry			
		Point	Center Line	Perpendicular Line	Glide Reflection
F_1	LLLLLLL				
F_2	NNNNNNN	×			
F_3	DDDDDDD		×		×
F_4	VVVVVVV			×	
F_5	HHHHHHH	×	×	×	×
F_6	ΛVΛVΛVΛ	×		×	×
F_7	LΓLΓLΓL				×

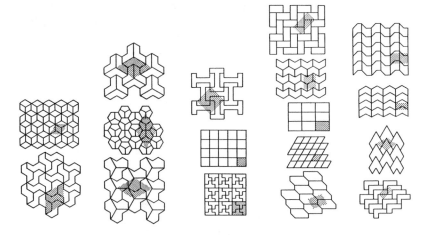

Figure 3.16

translation, there are patterns that remain invariant under a group generated by two translations along intersecting lines. These patterns are known as the *wallpaper* patterns. Figure 3.16, reprinted from Martin (1982b, p. 109), shows an example of each of the 17 possible wallpaper patterns. A detailed description of these patterns and their symmetry groups is contained in Chapter 11 of that text.

EXERCISES

1. Prove Theorem 3.34.

2. Explain why translations and glide reflections will not keep a polygon invariant.

3. Verify the entries in Table 3.1. [*Hint:* Use Theorem 3.28 to rewrite the half-turn as a product of reflections.]

4. Prove: If H_C is a half-turn with center C and $H_C(A) = B$, then C is the midpoint of $\overline{AB}$.

5. (a) Prove: If m and n are lines of symmetry for a figure α and m and n intersect at a point C, then there is a rotation with center C in the symmetry group for α. (b) Explain why a finite geometric figure cannot have two parallel lines of symmetry. (c) Can a finite geometric figure have two distinct points of symmetry? Why?

6. Verify the entries in Table 3.2 (see hint for Exercise 3).

7. (a) Show that each of the symmetries of an equilateral triangle can be expressed as the product of symmetries R_m and R_{120}. (b) What other pairs of symmetries of an equilateral triangle generate its symmetry group?

8. (a) Sketch a square and all of its points and lines of symmetry. Label each of these points and lines. (b) Determine the set of symmetries for this square. (c) Construct a multiplication table for this set of symmetries. (d) Use the table you constructed in part (c) to verify that the set of symmetries of a square is a group. What is the order of this group? What are the generators of this group?

9. Let G be a finite group of isometries that contains only rotations. Assume one element of G is a nonidentity rotation with center C. (a) Show that G cannot also contain a nonidentity rotation with center D where $D \neq C$. (*Hint*: Assume $R_{C,\theta}$ and $R_{D,\phi}$ are both elements of G and use Exercise 18 in Section 3.5 after showing that the isometry $(R_{D,\phi})^{-1}(R_{C,\theta})^{-1}R_{D,\phi}R_{C,\theta}$ is in G.) (b) Show that G is generated by a rotation $R_{C,\alpha}$ where α is the minimum positive angle of the rotations in G.

10. Let G be a finite group containing a reflection. (a) Prove that the subset of even isometries of G is a cyclic subgroup generated by a rotation $R_{C,\theta}$. [*Hint*: See Exercise 9.] (b) Prove: If R_m is a reflection in G, then $G = \langle R_{C,\theta}, R_m \rangle$.

11. Prove: (a) If T is a symmetry for a set of points α and P is a point of symmetry of α, then $T(P)$ is also a point of symmetry of α. (b) If T is a symmetry for a set of points α and l is a line of symmetry of α, then $T(l)$ is also a line of symmetry of α.

12. Prove: If τ is the translation that generates the cyclic subgroup of a frieze group, then τ is the shortest nonidentity translation of the group.

13. (a) Show that the only rotations in a frieze group with center c are the half-turns with centers on c. [*Hint*: See Exercise 20 in Section 3.5] (b) Show that the only translations in a frieze group with center c are those that keep c invariant.

Exercises 14–18 ask you to supply the reasons for some of the steps used in classifying the frieze groups. You will need to refer to the explanation of the notation given preceding the listing of the frieze groups.

14. (a) Prove that $\tau^n H_P$ is a half-turn. (b) Verify that the center of $\tau^{2k}H_P$ is P_k. (c) Verify that the center of $\tau^{2k+1}H_P$ is M_k.

15. Show that if A and B are points on line c, then $H_B H_A$ is a translation along c.

16. Let T be a translation along c and R_c a reflection with axis c. (a) Prove: $R_c T R_c = T$. (b) Use part (a) to verify that $R_c T = T R_c$.

17. Let T be a translation along c and R_p a reflection with axis p where p is perpendicular to c. (a) Prove: $T R_p T = R_p$. (b) Use part (a) to verify that $R_p T = T^{-1} R_p$.

18. (a) Prove that $\tau^n R_p$ is a reflection in a line perpendicular to c. (b) Verify that

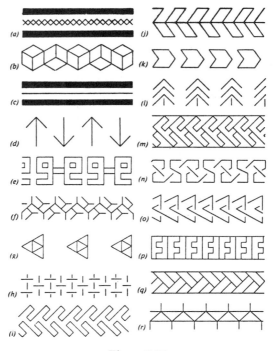

Figure 3.17

the axis of the reflection $\tau^{2k}R_p$ is incident with P_k. (c) Verify that the axis of the reflection $\tau^{2k+1}R_p$ is incident with M_k.

19. Make two frieze patterns for each of the seven frieze groups.

20. Name the frieze groups for each of the frieze patterns in Fig. 3.17, reprinted from Martin (1982b, p. 84).

3.8. Similarity Transformations

In previous sections of Chapter 3, we studied Euclidean geometry by exploring the invariant properties of V^* under the group of distance preserving transformations known as isometries. In this section we will determine which properties of V^* remain invariant under one-to-one linear transformations that preserve ratios of distance. The geometry determined by these transformations is called *similarity geometry*.

Definition 3.25. A *similarity with ratio r* is a one-to-one linear transformation T of V^* onto itself such that for each pair of points, P, Q $d^*(T(P), T(Q)) = rd^*(P, Q)$ for some nonzero real number r where d^* denotes directed distance.

Clearly, every isometry is a similarity with ratio ± 1, and thus isometries have all the properties of similarities. The converse is not true. However, since similarities are one-to-one linear transformations of V^* they also have 3×3 matrix representations with corresponding point and line equations $X' = AX$ and $ku' = uA^{-1}$, respectively (see Section 3.3). The form of the matrix of a similarity can be obtained by a method analogous to that used in the proof of Theorem 3.7. As in the case of isometries, there are direct and indirect similarities and the set of all similarities forms a group.

Theorem 3.35. *A similarity with ratio* r *has one of the following matrix representations:*

$$\begin{bmatrix} a_{11} & a_{12} & a_{13} \\ -a_{12} & a_{11} & a_{23} \\ 0 & 0 & 1 \end{bmatrix} \quad or \quad \begin{bmatrix} a_{11} & a_{12} & a_{13} \\ a_{12} & -a_{11} & a_{23} \\ 0 & 0 & 1 \end{bmatrix} \quad where \ a_{11}^2 + a_{12}^2 = r^2.$$

Theorem 3.36. *The set of similarities forms a group of which the set of isometries is a subgroup.*

Figures that correspond to each other under a similarity are said to be *similar*. The verification that similar triangles do indeed have angles of the same measure and sides of proportional measure is nearly a replication of the proofs of comparable theorems for congruent triangles (see Section 3.4).

Definition 3.26. Two sets of points α and β are *similar*, denoted $\alpha \sim \beta$, if β is the image of α under a similarity.

Theorem 3.37. *Let* u' *and* v' *be the images of lines* u *and* v *under a similarity. If the similarity is direct then* $m(\angle(u'v')) = m(\angle(u, v))$. *If the similarity is indirect then* $m(\angle(u', v')) = -m(\angle(u, v))$.

Theorem 3.38. *If* $\triangle PQR \sim \triangle P'Q'R'$ *then there exists an* $r \neq 0$ *such that* $m(\overline{P'Q'}) = r(m(\overline{PQ}))$, $m(\overline{Q'R'}) = r(m(\overline{QR}))$, $m(\overline{R'P'}) = r(m(\overline{RP}))$, $m(\angle P'Q'R') = \pm m(\angle PQR)$, $m(\angle Q'R'P') = \pm m(\angle QRP)$, *and* $m(\angle R'P'Q') = \pm m(\angle RPQ)$.

To verify the converse of this last theorem, it is necessary to determine more about the behavior of similarities. Fortunately, we need only consider one particular type of similarity.

Definition 3.27. Let C be an arbitary point and r a nonzero real number. A *dilation with center* C *and ratio* r, denoted $D_{C,r}$, is a direct similarity with ratio r and invariant point C that maps any point P to a point P' on line CP (see Fig. 3.18). Dilations are also called dilatations or central similarities.

Using this definition, the invariant points and lines of a dilation can be determined (see Exercise 9) and the matrix representation can be found.

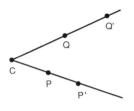

Figure 3.18

Theorem 3.39. *Under a dilation $D_{C,r}$, the point C and each line incident with C are invariant.*

Theorem 3.40. *A dilation with center $O(0,0,1)$ and ratio r has matrix representation*

$$\begin{bmatrix} r & 0 & 0 \\ 0 & r & 0 \\ 0 & 0 & 1 \end{bmatrix}$$

A dilation with center $C(c_1, c_2, 1)$ has matrix representation

$$\begin{bmatrix} r & 0 & c_1(1-r) \\ 0 & r & c_2(1-r) \\ 0 & 0 & 1 \end{bmatrix}$$

Proof (Outline). For the first case, the requirement that a direct similarity with matrix $A = [a_{ij}]$ keep $O(0,0,1)$ invariant implies that $a_{13} = a_{23} = 0$. The requirement that any point $X(x,0,1)$ on the line $[0,1,0]$ must map to a point $X'(x',0,1)$ also on this line yields $a_{12} = 0$ and $a_{11} = r$.

The second case can be verified after noting that $D_{C,r} = T D_{O,r} T^{-1}$ where T is the translation mapping O to C. $\qquad \square$

Using this matrix representation we can now determine the effect of a dilation on lines that are not incident with its center.

Theorem 3.41. *If $D_{C,r}$ is a dilation with $r \neq 1$ and m is a line not incident with C, then $D_{C,r}(m) = m'$ is a distinct line parallel to m.*

Proof. The line equation of this dilation requires the matrix of $(D_{C,r})^{-1}$. Since this transformation is also a dilation with center C and ratio $r' = 1/r$ (see Exercise 11), its matrix representation is given by Theorem 3.40.

Using this matrix in the line equation of the dilation, we can find m', the image of the line $m[m_1, m_2, m_3]$ as follows:

$$[m_1, m_2, m_3] \begin{bmatrix} r' & 0 & c_1(1-r') \\ 0 & r' & c_2(1-r') \\ 0 & 0 & 1 \end{bmatrix} = [r'm_1, r'm_2, m_3']$$

where $m_3' = m_1 c_1(1-r') + m_2 c_2(1-r') + m_3$.

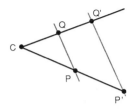

Figure 3.19

Clearly, $m' = [r'm_1, r'm_2, m'_3]$ is equal to m iff $m'_3 = r'm_3$, that is, iff

$$m_1c_1(1 - r') + m_2c_2(1 - r') + m_3 = r'm_3$$

or

$$m_1c_1(1 - r') + m_2c_2(1 - r') + m_3(1 - r') = 0$$

or

$$m_1c_1 + m_2c_2 + m_3 = 0 \quad \text{since } r' \neq 1;$$

but this is exactly the condition that makes m incident with C. Thus if m is not incident with C, m' is necessarily a distinct line parallel to m. $\square$

Given the center C, points P and P' (P' on line CP), it is now possible to construct the image of any other point under a dilation $D_{C,r}$ that maps P to P'. This construction demonstrates that a dilation is uniquely determined by its center together with point and its image.

Case 1. Q is not on CP (see Fig. 3.19). Q' will be the point of intersection of line CQ and the line through P' parallel to PQ.

Case 2. Q is on CP (see Fig. 3.20). We can find the image of a point R not on CP as before and then use R and R' in place of P and P' in case 1. The matrix representation of dilations can be used to characterize the similarities in terms of dilations and isometries (see Exercise 13).

Theorem 3.42. *Every similarity can be expressed as the product of a dilation and an isometry.*

With this characterization, it is possible to outline a proof of the converse of Theorem 3.38 similar to the outline given for Theorem 3.23 (see Exercise 15).

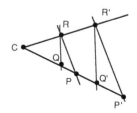

Figure 3.20

Theorem 3.43. *If $\triangle PQR$ and $\triangle P'Q'R'$ are two triangles with $m(\overline{P'Q'}) = r(m(\overline{PQ}))$, $m(\overline{Q'R'}) = r(m(\overline{QR}))$, $m(\overline{R'P'}) = r(m(\overline{RP}))$, and also $m(\angle P'Q'R') = \pm m(\angle PQR)$, $m(\angle Q'R'P') = \pm m(\angle QRP)$, and $m(\angle R'P'Q') = \pm m(\angle RPQ)$, then there is a similarity mapping $\triangle PQR$ to $\triangle P'Q'R'$.*

Another major result that follows directly from the proof of Theorem 3.43 has to do with the increased freedom allowed by similarities. Whereas an isometry can always be found to map a point P to an arbitrary image point P', it is not possible to use an isometry to map a pair of points P, and Q to a second arbitrary pair of points P' and Q' (why not?). Such a mapping can be accomplished with a similarity (see Exercise 17).

Theorem 3.44. *There exist two similarities, one direct and one indirect, which map a pair of distinct points P and Q to a pair of corresponding points P' and Q'.*

EXERCISES

1. Show that similarities preserve betweenness (see Definition 3.13).

2. Prove: If a similarity maps $\overline{PQ}$ and $\overline{RS}$ to $\overline{P'Q'}$ and $\overline{R'S'}$, respectively, and $d(P, Q) = s(d(R, S))$, then $d(P'Q') = s(d(R', S'))$. (This proves that similarities preserve ratios of distance.)

3. Prove Theorem 3.35.

4. Prove Theorem 3.36

5. Prove Theorem 3.37.

6. Prove Theorem 3.38.

7. Let C, P, and P' be points with coordinates, $C(3, -2, 1)$, $P(1, 0, 1)$, and $P'(7, -6, 1)$. (a) Show that these three points are collinear. (b) Find the matrix of a dilation with center C that maps P to P'. (c) Find the image of lines $m[1, 1, -1]$ and $n[1, 1, 1]$ under this dilation.

8. Show that a rotation with angle $180°$ is a dilation.

9. Prove Theorem 3.39.

10. Complete the proof of Theorem 3.40 outlined in the text.

11. Show $(D_{C,r})^{-1} = D_{C,1/r}$.

12. Prove that the *only* invariant points and lines under a nonidentity dilation with center C are C and lines through C.

13. Prove Theorem 3.42.

14. Find the product of a translation, rotation, and dilation that maps $\triangle PQR$ to $\triangle P'Q'R'$ where $P(3, 6, 1)$, $Q(-2, 5, 1)$, $R(-3, -1, 1)$, $P'(0, 0, 1)$, $Q'(2, -10, 1)$, $R'(14, -12, 1)$. [*Hint*: translate P to P' first.]

15. Outline a proof of Theorem 3.43.

16. Find matrices of two different similarities both of which map $P(1, 2, 1)$ and $Q(0, 0, 1)$ to $P'(2, 4, 1)$ and $Q'(-4, 2, 1)$, respectively. What is the image of $R(1, 1, 1)$ under each?

17. Prove Theorem 3.44.

3.9. Affine Transformations

In Section 3.8, similarities were shown to be generalizations of isometries. In this section we will continue this process of generalizing the isometries by considering the unrestricted set of one-to-one linear transformations of V^*. The geometry determined by these transformations is called *affine geometry*.

Definition 3.28. An *affinity* is a one-to-one linear transformation of V^* onto itself.

In other words, the affinities are the transformations described in Section 3.3. There we discovered that affinities map points according to the matrix equation $X' = AX$ where

$$A = \begin{bmatrix} a_{11} & a_{12} & a_{13} \\ a_{21} & a_{22} & a_{23} \\ 0 & 0 & 1 \end{bmatrix} \quad \text{and} \quad |A| \neq 0$$

We also noted that affinities preserve collinearity and map lines according to the matrix equation $ku' = uA^{-1}$. Theorem 3.4 can be reworded to state that the set of affinities form a group and Theorem 3.36 in Section 3.8 implies that the set of similarities forms a subgroup of the group of affinities.

Since similarities and isometries are specific types of affinities, any properties invariant under affinities are also invariant under similarities and isometries. One of the most important of these invariant properties is parallelism.

Theorem 3.45. *If T is an affinity and m and n are parallel lines, then $T(m)$ is parallel to $T(n)$.*

Proof. Assume $m[m_1, m_2, m_3]$ and $n[n_1, n_2, n_3]$ are parallel lines. Then there is a nonzero real number t such that $n_1 = tm_1$ and $n_2 = tm_2$. We can find $T(m)$, the image of line m under the affinity T with matrix A, using the line equation $k_1 m' = mB$ where $B = A^{-1}$. Specifically,

$$k_1 m' = [m_1, m_2, m_3] \begin{bmatrix} b_{11} & b_{12} & b_{13} \\ b_{21} & b_{22} & b_{23} \\ 0 & 0 & 1 \end{bmatrix}$$

or

$$m'[m'_1, m'_2, m'_3] = (1/k_1)[m_1 b_{11} + m_2 b_{21}, m_1 b_{12} + m_2 b_{22}, m_1 b_{13} + m_2 b_{23} + m_3]$$

Similarly,

$$n'[n'_1, n'_2, n'_3] = (1/k_2)[n_1 b_{11} + n_2 b_{21}, n_1 b_{12} + n_2 b_{22}, n_1 b_{13} + n_2 b_{23} + n_3]$$

Then, substituting $n_1 = tm_1$ and $n_2 = tm_2$ yields

$$n' = (1/k_2)[tm_1, b_{11} + tm_2 b_{21}, tm_1 b_{12} + tm_2 b_{22}, n_1 b_{13} + n_2 b_{23} + n_3]$$

or

$$n'_i = (t/k_2)(m_1 b_{1i} + m_2 b_{2i}) = (tk_1/k_2)m'_i \quad \text{for } i = 1, 2$$

Thus m' and n' are parallel.   □

Clearly, general affinities do not preserve distance as do isometries, nor do they preserve ratios of distances as do similarities. However, they do preserve a more general ratio of distances known as a *segment division ratio*. This is verified by the proof of the next theorem (see Exercise 1).

Theorem 3.46. *If T is an affinity and P, Q, and R are three distinct collinear points such that*

$$\frac{d(Q, P)}{d(Q, R)} = k, \quad \text{then} \quad \frac{d(T(Q), T(P))}{d(T(Q), T(R))} = k$$

We can use this theorem along with Definition 3.13 to show that affinities preserve betweenness of points. It then follows that affinities also preserve segments and their midpoints.

Theorem 3.47. *If T is an affinity and P, Q, and R are three collinear points with P, between Q and R, then T(P) is between T(Q) and T(R).*

Proof. Since P is between Q and R, $d(Q, P) + d(P, R) = d(Q, R)$. Dividing each term of this equation by $d(Q, R)$ and letting $d(Q, P)/d(Q, R) = k$, gives $d(P, R)/d(Q, R) = 1 - k$. By Theorem 3.46, $d(Q', P')/d(Q', R') = k$ and $d(P', R')/d(Q', R') = 1 - k$ where $P' = T(P)$, and so on. Substitution then yields $d(P', R')/d(Q', R') = 1 - d(Q', P')/d(Q', R')$ or $d(Q', P') + d(P', R') = d(Q', R')$, so P' is between Q' and R'.   □

Corollary. *If T is an affinity and M is the midpoint of the segment with endpoints Q and R, then T(M) is the midpoint of the segment with endpoints T(Q) and T(R).*

Proof. Merely let $k = \frac{1}{2}$ in the proof of Theorem 3.47.   □

We can gain an intuitive understanding of the effect of affinities by considering two specific types known by the suggestive names of *shears* and *strains*.

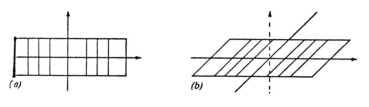

Figure 3.21

Definition 3.29. A *shear with axis* m, denoted S_m, is an affinity that keeps m pointwise invariant and maps every other point P to a point P' so that the line PP' is parallel to m (see Fig. 3.21).

Theorem 3.48. *The matrix representation of a shear with axis* $x[0, 1, 0]$ *is*

$$\begin{bmatrix} 1 & j & 0 \\ 0 & 1 & 0 \\ 0 & 0 & 1 \end{bmatrix}$$

In general the matrix representation of a shear S_m *can be found using* $S_m = SS_xS^{-1}$ *where* S *is a direct isometry mapping* x *to* m $(S(x) = m)$.

Proof. Since S_x keeps each point on the line $x[0, 1, 0]$ invariant, the following equation must be true for all real numbers x_1

$$\begin{bmatrix} a_{11} & a_{12} & a_{13} \\ a_{21} & a_{22} & a_{23} \\ 0 & 0 & 1 \end{bmatrix} \begin{bmatrix} x_1 \\ 0 \\ 1 \end{bmatrix} = \begin{bmatrix} x_1 \\ 0 \\ 1 \end{bmatrix}$$

Therefore $a_{11}x_1 + a_{13} = x_1$ and $a_{21}x_1 + a_{23} = 0$, yielding $a_{11} = 1$, $a_{13} = 0$, $a_{21} = 0$, and $a_{23} = 0$. If $P(p_1, p_2, 1)$ is a point not on line $x[0, 1, 0]$ (so $p_2 \neq 0$), P must map to a point P' on the line through P parallel to line x. This line has coordinates $u[0, 1, -p_2]$, so P' must have coordinates $P'(p'_1, p_2, 1)$ leading to the following equation

$$\begin{bmatrix} 1 & a_{12} & 0 \\ 0 & a_{22} & 0 \\ 0 & 0 & 1 \end{bmatrix} \begin{bmatrix} p_1 \\ p_2 \\ 1 \end{bmatrix} = \begin{bmatrix} p'_1 \\ p_2 \\ 1 \end{bmatrix}$$

so $a_{22}p_2 = p_2$. Since $p_2 \neq 0$, this implies that $a_{22} = 1$. Therefore the matrix does have the form given in the statement of the theorem.

The verification of the second part of the theorem is analogous to that used to prove similar results in previous theorems.   □

Strains are also defined in terms of a pointwise invariant line and the procedure used to determine their matrix representation is similar to that just used to find the matrix of a shear (see Exercise 4).

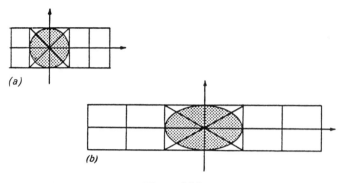

(a)

(b)

Figure 3.22

Definition 3.30. A *strain with axis m*, denoted T_m, keeps m pointwise invariant and maps every other point P to a point P' so that the line PP' is perpendicular to m (see Fig. 3.22).

Theorem 3.49. *The matrix representation of a strain with axis* $x[0, 1, 0]$ *is*

$$\begin{bmatrix} 1 & 0 & 0 \\ 0 & k & 0 \\ 0 & 0 & 1 \end{bmatrix}$$

In general, the matrix representation of strain T_m *can be found using* $T_m = ST_x S^{-1}$ *where S is a direct isometry mapping x to m* ($S(x) = m$).

Using shears and strains along with similarities it is possible to obtain any affinity. In particular, we can obtain any affinity as a product of a shear S_x, a strain T_x, and a direct similarity.

Theorem 3.50. *Any affinity can be written as the product of a shear, a strain, and a direct similarity.*

Proof. We can verify this theorem by merely demonstrating that the following product does indeed yield the matrix of a general affinity as indicated:

$$\begin{bmatrix} a_{11} & a_{12} & a_{13} \\ a_{21} & a_{22} & a_{23} \\ 0 & 0 & 1 \end{bmatrix} = \begin{bmatrix} a_{11} & -a_{21} & a_{13} \\ a_{21} & a_{11} & a_{23} \\ 0 & 0 & 1 \end{bmatrix} \begin{bmatrix} 1 & 0 & 0 \\ 0 & k & 0 \\ 0 & 0 & 1 \end{bmatrix} \begin{bmatrix} 1 & j & 0 \\ 0 & 1 & 0 \\ 0 & 0 & 1 \end{bmatrix}$$

where

$$j = \frac{(a_{11}a_{12} + a_{21}a_{22})}{(a_{11}^2 + a_{21}^2)} \quad \text{and} \quad k = \frac{(a_{11}a_{22} - a_{12}a_{21})}{(a_{11}^2 + a_{21}^2)}.$$

$\square$

As indicated in the previous section, the more general the transformations become, the more freedom they allow. Whereas isometries exist that map a point P to a point P', and similarities exist that map a pair of points P, Q to a pair of points P', Q', the next theorem shows that affinities exist that map three noncollinear points P, Q, R to three noncollinear points P', Q', R'.

Theorem 3.51. *Given two triangles, $\triangle PQR$ and $\triangle P'Q'R'$, there is an affinity mapping $\triangle PQR$ to $\triangle P'Q'R'$.*

Proof. We can show that there is an affinity mapping P, Q, and R to P', Q', and R', resectively, by finding a matrix A such that $P' = AP$, $Q' = AQ$, and $R' = AR$. This involves six equations in six unknowns. However, in actual practice we can simpify the determination of the matrix A as follows: First find the matrix of the affinity S that maps $O(0,0,1)$, $X(1,0,1)$, and $U(1,1,1)$ to P, Q, and R, respectively. Then find the matrix of the affinity T, which maps O, X, and U to P', Q', and R'. The affinity TS^{-1} will map P, Q, and R to P', Q', and R'. Since affinities preserve betweenness, TS^{-1} also maps the segments $\overline{PQ}$, $\overline{QR}$, and $\overline{RP}$ to $\overline{P'Q'}$, $\overline{Q'R'}$, and $\overline{R'P'}$, and therefore $\triangle PQR$ to $\triangle P'Q'R'$.   □

In addition to segments and triangles, affinities also preserve other geometric figures. Since isometries preserve distance and each of the conic sections (circles, ellipses, parabolas, and hyperbolas) can be characterized in terms of distances, it is obvious that isometries preserve each of the conic sections, for example, the image of a circle under an isometry is a circle. To explore the invariance of conic sections under more general linear transformations, it is convenient to note that all conic sectons can be written via matrix equations (see Exercise 10).

Theorem 3.52. *Any conic section can be written algebraically as*

$$c_{11}x_1^2 + c_{22}x_2^2 + 2c_{13}x_1 + 2c_{23}x_2 + 2c_{12}x_1x_2 + c_{33} = 0$$

or, in matrix notation, as

$$[x_1, x_2, 1]\begin{bmatrix} c_{11} & c_{12} & c_{13} \\ c_{12} & c_{22} & c_{23} \\ c_{13} & c_{23} & c_{33} \end{bmatrix}\begin{bmatrix} x_1 \\ x_2 \\ 1 \end{bmatrix} = 0 \quad \text{or} \quad X^t C X = 0.$$

The symmetric matrix $C = [c_{ij}]$ is called the matrix of the conic section. The conic section is nondegenerate (i.e., it is not a line, pair of lines, point, or the empty set iff $|C| \neq 0$). Furthermore, a conic section is an ellipse, hyperbola, or parabola if $(c_{12})^2 - c_{11}c_{22} < 0$, $(c_{12})^2 - c_{11}c_{22} > 0$, or $(c_{12})^2 - c_{11}c_{22} = 0$ (but c_{11} and c_{22} are not *both* zero), respectively. Thus there are three distinct types of conic sections where circles ($c_{11} = c_{22}$) are considered to be special cases of ellipses.

Using matrix notation, it is relatively easy to determine the matrix of the image of a conic section under an affinity. The entries in this second matrix show that affinities preserve types of conic sections.

Theorem 3.53. *The image of a conic section under an affinity is a conic section of the same type. Furthermore, if A is the matrix of an affinity, then the matrix of the image conic section is* $C' = (A^{-1})^t C A^{-1}$.

Proof. Under the affinity, X is mapped to $X' = AX$. Solving for X, gives $X = A^{-1}X'$. Substituting this into the matrix equation $X^t C X = 0$ yields $(A^{-1}X')^t C(A^{-1}X') = 0$ or $X'^t((A^{-1})^t C A^{-1})X' = 0$. This latter equation is the equation of a conic section with symmetric matrix $C' = (A^{-1})^t C A^{-1}$, where $|C'| = 0$ iff $|C| = 0$. To show that the type of conic section is preserved requires a straightforward but somewhat tedious calculation. $\qquad\qquad\square$

EXERCISES

1. Prove Theorem 3.46 for the case where P, Q, and R have coordinates $P(x, 0, 1)$, $Q(0, 0, 1)$, $R(y, 0, 1)$.

2. Find the matrix of a shear with axis $x_1 = x_2$.

3. Find the matrix of a strain with axis $x_1 = 5$.

4. Prove Theorem 3.49. [*Hint*: See the proof of Theorem 3.48.]

5. Show that a dilation with center O is the product of strains with axes $x[0, 1, 0]$ and $y[1, 0, 0]$.

6. Find the matrix of an affinity mapping $P(1, -1, 1)$, $Q(2, 1, 1)$, and $R(3, 0, 1)$ to $P'(0, 1, 1)$, $Q'(1, 2, 1)$, and $R'(0, 3, 1)$, respectively. [*Hint*: Use the method described in the proof of Theorem 3.51.]

7. Show that the only affinity with three noncollinear invariant points is the identity. [*Hint*: First assume the invariant points are $O(0, 0, 1)$, $X(1, 0, 1)$, and $U(1, 1, 1)$.]

8. Use Exercise 7 to show that there is a *unique* affinity mapping any three noncollinear points to any three noncollinear points. [*Hint*: Assume S and T are two such affinities and consider the affinity ST^{-1}.]

9. Show that affinities preserve parallelograms.

10. Verify that the standard equation for a conic section given in Theorem 3.52 is equivalent to the given matrix equation.

11. Let $A = \begin{bmatrix} 3 & -2 & 4 \\ 2 & -1 & -2 \\ 0 & 0 & 1 \end{bmatrix}$.

    Find the image of the parabola $y = 6x^2$ under the affinity with matrix A. Verify that the image is also a parabola.

12. Show that the image of a circle under a similarity is again a circle. (Note: In general, affinities can map circles to noncircular ellipses.)

Exercises 13 and 14 require the use of the following linear algebra formula for the area of a triangle with vertices $P(p_1, p_2, 1)$ $Q(q_1, q_2, 1)$, and $R(r_1, r_2, 1)$:

$$\text{area}(\triangle PQR) = \tfrac{1}{2}\text{abs}\left[\det\begin{bmatrix} p_1 & q_1 & r_1 \\ p_2 & q_2 & r_2 \\ 1 & 1 & 1 \end{bmatrix}\right]$$

13. Prove: If T is an affinity with matrix A, and T maps $\triangle PQR$ to $\triangle P'Q'R'$, then $\text{area}(\triangle P'Q'R') = k(\text{area}(\triangle PQR))$ where $k = \text{abs}(\det(A))$. (Note: When $k = 1$, the affinity is called an *equiareal* transformation.)

14. Using Exercise 13 show that the area of a triangle is preserved under isometries and shears.

3.10. Suggestions for Further Reading

Coxford, A.F., and Usiskin, Z.P. (1971). *Geometry: A Transformation Approach.* River Forest, IL: Laidlow Brothers. Uses transformations in its presentation of the standard topics of elementary Euclidean geometry.

Dodge, C.W. (1972). *Euclidean Geometry and Transformations.* Reading, MA: Addison-Wesley. Chapters 2 and 3 contain an elementary presentation of isometries and similarities and include applications.

Eccles, F.M. (1971). *An Introduction to Transformational Geometry.* Menlo Park, CA: Addison-Wesley. Intended to introduce high-school students to the transformations following a traditional geometry course.

Gans, D. (1969). *Transformations and Geometries.* New York: Appleton-Century-Crofts. A detailed presentation of the transformations introduced in this chapter followed by a presentation of the more general projective and topological transformations.

Iaglom, I.M. (1962). *Geometric Transformations,* Vols. 1, 2, 3. New York: Random House. Numerous problems of elementary Euclidean geometry are solved through transformations.

Jeger, M. (1969). *Transformation Geometry.* London: Allen and Unwin. Numerous diagrams are included in this easy-to-understand presentation of isometries, similarities, and affinities.

Martin, G.E. (1982b). *Transformation Geometry: An Introduction to Symmetry.* New York: Springer-Verlag. Introduces isometries and applies them to ornamental groups and tessellations.

Maxwell, E.A. (1975). *Geometry by Transformations.* Cambridge: Cambridge University Press. A high-school-level introduction of isometries and similarities including their matrix representations.

Readings on Tiling the Plane and Paper Folding

Faulkner, J.E. (1975). Paper folding as a technique in visualizing a certain class of transformations. *Mathematics Teacher* 68: 376–377.

Gardner, M. (1975). On tessellating the plane with convex polygon tiles. *Scientific American* 233(1): 112–117.

Gardner, M. (1978). The art of M.C. Escher. In: M. Gardner, *Mathematical Carnival,* pp. 89–102. New York: Alfred A. Knopf.

Grünbaum, B., and Shephard, G.C. (1987). *Tilings and Patterns*. New York: W.H. Freeman.

Haak, S. (1976). Transformation geometry and the artwork of M.C. Escher. *Mathematics Teacher* 69: 647–652.

Johnson, D.A. (1973). *Paper Folding for the Mathematics Class*. Reston, VA: N.C.T.M.

MacGillavry, C.H. (1976). *Symmetry Aspects of M.C. Escher's Periodic Drawings*, 2d ed. Utrecht: Bohn, Scheltema & Holkema.

O'Daffer, P.G., and Clemens, S.R. (1976). *Geometry: An Investigative Approach*. Menlo Park, CA: Addison-Wesley.

Olson, A.T. (1975). *Mathematics Through Paper Folding*. Reston, VA: N.C.T.M.

Ranucci, E.R. (1974). Master of tessellations: M.C. Escher, 1898–1972. *Mathematics Teacher* 67: 299–306.

Robertson, J. (1986). Geometric constructions using hinged mirrors. *Mathematics Teacher* 79: 380–386.

Teeters, J.C. (1974). How to draw tessellations of the Escher type. *Mathematics Teacher* 67: 307–310.

Suggestions for Viewing

Adventures in Perception (1973, 22 min). An especially effective presentation of the work of M.C. Escher. Produced by Hans Van Gelder, Film Producktie, N.V., The Netherlands. Available from Phoenix/B.F.A. Films, 468 Park Ave. S., New York, NY 10016 (800-221-1274).

Dihedral Kaleidoscopes (1971; 13 min). Uses pairs of intersecting mirrors (dihedral kaleidoscopes) to demonstrate several regular figures and their stellations and tilings of the plane. Produced by the College Geometry Project at the University of Minnesota. Available from International Film Bureau, 332 South Michigan Ave., Chicago, IL 60604.

Isometries (1971; 26 min). Demonstrates that every plane isometry is a translation, rotation, reflection, or glide reflection and that each is the product of at most three reflections. Produced by the College Geometry Project at the University of Minnesota. Available from International Film Bureau, 332 South Michigan Ave., Chicago, IL 60604.

Symmetries of the Cube (1971; 13.5 min). Uses mirrors to exhibit the symmetries of a square as a prelude to the analogous generation of the cube by reflections. Produced by the College Geometry Project at the University of Minnesota. Available from International Film Bureau, 332 South Michigan Ave., Chicago, IL 60604.

CHAPTER 4

Projective Geometry

4.1. Gaining Perspective

From the analytic viewpoint of Klein's definition of geometry, projective geometry is the logical generalization of the affine geometry introduced in Chapter 3. Just as we were able to generalize the isometries of the Euclidean plane to similarities, and these in turn to affinities, we will now be able to generalize affinities to collineations, the transformations that define projective geometry. There is, however, one new ingredient required in this last generalization. The set of points contained in the Euclidean plane must be enlarged to include points on one additional line, a line often referred to as the *ideal* line. Rather than complicating the geometry, these new ideal points simplify projective geometry and give it the highly desirable property of duality.

The historical development of projective geometry, however, was synthetic, rather than analytic, in nature. The origins of this geometry can be traced to the attempts of Renaissance painters to achieve realistic representations of three-dimensional objects on two-dimensional canvas. These painters, influenced by Plato's thesis that nature is mathematically designed, sought and found mathematical relations that could be used to achieve perspective. This interplay of mathematics and art, the importance of Plato's thesis, and the influence of the church make the origins of projective geometry a fascinating episode in the history of mathematics. This history is detailed in the sources given at the end of this chapter. These readings should explain the connection between the *ideal* points referred to at the beginning of this section and the *vanishing* points used in paintings.

The relevance of projective geometry to achieving realistic planar representations of three-dimensional objects is currently making the study of projective geometry a prerequisite to the study of computer graphics. The value of this prerequisite is enhanced, since computer graphics uses the analytic representations of points and lines by homogeneous coordinates and the representation of transformations by matrices developed in projective geometry.

4.2. The Axiomatic System and Duality

Before introducing an analytic model for plane projective geometry, it is necessary to develop an axiomatic system for this geometry. The axiom system we will consider contains six axioms, however, we will call any system satisfying the first four of these axioms a *projective plane*.* It is these first four axioms we will consider in this section as we begin our synthetic treatment of plane projective geometry. Just as in the axiom systems of Chapter 1, the undefined terms for this system are "point," "line," and "incidence"; points are said to be *collinear* if they are incident with the same line. The term "complete quadrangle" used in Axiom 4 will be explained in Definition 4.2.

Axioms for a Projective Plane

Axiom 1. Any two distinct points are incident with exactly one line.
Axiom 2. Any two distinct lines are incident with at least one point.
Axiom 3. There exist at least four points, no three of which are collinear.
Axiom 4. The three diagonal points of a complete quadrangle are never collinear.

Note that although the first axiom is characteristic of Euclidean geometry, the second axiom, guaranteeing that pairs of lines intersect, is not; that is, there do not exist parallel lines in this geometry. Also notice that Axioms 1 and 2 are nearly dual statements. (Recall that the dual of a statement is obtained by replacing each occurrence of the word "point" by the word "line" and vice versa.) The dual of Axiom 1 would read: "Any two distinct lines are incident with exactly one point." A proof of this statement follows trivially from Axioms 1 and 2 (see Exercise 1), and thus the duals of both axioms are theorems of this axiomatic system.

A careful reading will show that Axioms 1 and 2 do *not* assert the existence of either points or lines. However, Axiom 3 and its dual assure us that points and lines do exist in the projective plane.

Theorem 4.1 (Dual of Axiom 3). *There exist at least four lines, no three of which are concurrent.*

Proof. Let *A*, *B*, *C*, *D* be four points, no three collinear, as guaranteed by Axiom 3. Then by Axiom 1, there exist the four lines *AB*, *AC*, *CD*, and *BD*. If any three of these were concurrent, the dual of Axiom 1 would be contradicted. □

---

*In general a *projective plane* is a system that satisfies Axioms 1–3 and an axiom that guarantees that every line contains at least three points.

As in the preceding proof, points of this geometry are denoted by capital letters, A, B, C, and so on, while lines are denoted by small letters a, b, c, and so on. A pair of capital letters, AB, refers to the unique line determined by points A and B. Since a pair of lines a and b also determines a unique point, we denote this point by $a \cdot b$. In addition, we use the notation AIa or aIA to indicate that point A and line a are incident.

Since Axiom 3 guarantees the existence of three noncollinear points, figures resembling Euclidean triangles exist. However, since there is no concept of betweenness in this geometry, the sides of a triangle are lines, *not* segments. This latter change makes the following definition self-dual.

Definition 4.1. A *triangle* is a set of three noncollinear points and the three lines determined by these points. The points are called *vertices* and the lines are called *sides* of the triangle (see Fig. 4.1).

Figures consisting of four points and the lines they determine also exist. Unlike triangles, these figures have no comparable analogues in Euclidean geometry.

Definition 4.2. A (*complete*) *quadrangle* is a set of four points, no three collinear, and the six lines determined by these four points. The points are called *vertices* and the lines are called *sides* of the quadrangle. If A, B, C, D are the four points of a quadrangle, then AB and CD, AC and BD, and AD and BC are said to be pairs of *opposite sides*. The points at which pairs of opposite sides intersect are called *diagonal points* of the quadrangle (see Fig. 4.2).

As asserted in Axiom 4, the diagonal points of a complete quadrangle form a triangle known as the *diagonal* triangle of the quadrangle. The existence of this

Figure 4.1

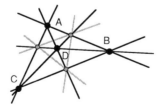

Figure 4.2

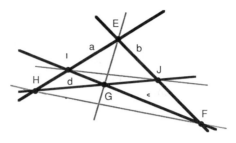

Figure 4.3

diagonal triangle can be used to show that each line in the projective plane contains at least four points (see Exercise 2).

To determine whether the dual of Axiom 4 is a theorem, it is necessary to consider the dual of Definition 4.2. Unlike the definition of a triangle, Definition 4.2 is not self-dual. Therefore the dual of a quadrangle is another figure of this geometry.

Definition 4.3. A (*complete*) *quadrilateral* is a set of four lines, no three concurrent, and the six points determined by these lines. The points are called *vertices* and the lines are called *sides* of the quadrilateral. If a, b, c, d are the four lines of the quadrilateral, $a \cdot b$ and $c \cdot d$, $a \cdot c$ and $b \cdot d$, and $a \cdot d$ and $b \cdot c$ are said to be pairs of *opposite vertices*. The lines joining pairs of opposite vertices are called *diagonal lines* of the quadrilateral (see Fig. 4.3).

Theorem 4.2 (Dual of Axiom 4). *The three diagonal lines of a complete quadrilateral are never concurrent.*

Proof. Let $abcd$ be an arbitrary complete quadrilateral. Let $E = a \cdot b$, $F = b \cdot c$, $G = c \cdot d$, $H = a \cdot d$, $I = a \cdot c$, and $J = b \cdot d$. Then the diagonal lines are EG, FH, and IJ. Assume these three lines are concurrent; that is, EG, FH, and IJ intersect at a point. But $EFGH$ forms a complete quadrangle with diagonal points $EF \cdot GH = b \cdot d = J$, $EG \cdot FH$, and $EH \cdot FG = a \cdot c = I$, but since EG, FH, and IJ are concurrent, this implies that the diagonal points of the complete quadrangle $EFGH$ are collinear, contradicting Axiom 4. Thus the diagonal lines of complete quadrilateral $abcd$ are not concurrent.   □

Hence the diagonal lines of a complete quadrilateral also determine a triangle known as the *diagonal triangle* of the quadrilateral.

With the proof of Theorem 4.2 we have completed the process of showing that the axiomatic system consisting of Axioms 1–4 satisfies the *principle of duality* (see Section 1.3). Eventually we shall add two more axioms to this system and verify that this larger system also satisfies the principle of duality.

Even though our objective in this chapter is the study of the real projective plane, it is interesting to note that Axioms 1–3 are essentially the same as three of the four axioms for finite projective planes given in Section 1.3. The remaining axiom for finite projective planes (Axiom P2) refers to the number of points on a line. As indicated previously in this section, Axiom 4 guarantees that there are at least four points on each line. Thus any finite model of our current axiomatic system is of order $n \geq 3$ and so contains at least 13 points. In fact, the 13-point model (model 3) given in Section 1.3 is also a model for Axioms 1–4. The verification that this model actually satisfies Axiom 4 consists of a tedious case-by-case check of all possible quadrangles (see Exercise 6).

An infinite model of this axiomatic system can be obtained by slightly extending a Euclidean plane as follows.

An Infinite Model for the Projective Plane

Let π be a plane parallel to, but not equal to the x–y plane in *Euclidean* 3-space and let O denote the origin of the Cartesian coordinate system. Note that each point P in π, together with the point O, determines a unique line p, so P can be said to correspond to a unique line through O, namely, the line p. Similarly, each line l in π, together with the point O, determines a unique plane λ, so l can be said to correspond to a plane through O, namely, λ (see Fig. 4.4). This correspondence is clearly a one-to-one mapping of the set of points and lines in π into the set of lines and planes through O. However, there is one plane, namely, the x–y plane and a subset of lines, namely, the set of all lines through O in the x–y plane that are "missed" by this mapping.

A model, π', of the projective plane is obtained by adding an *ideal* line and *ideal* points to π to make this correspondence not only one-to-one but also onto. The *ideal* line, which is added to π, corresponds to the x–y plane and the *ideal* points added to π correspond to those lines through O that lie in the x–y plane. Once added, this ideal line and these ideal points are considered to be indistinguishable from the other lines and points in π'.

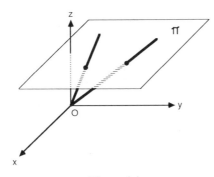

Figure 4.4

In addition to describing the points and lines of π' it is necessary to describe the interpretation of the term "incidence." A point and line in π' are said to be incident iff the corresponding line through O lies in the corresponding plane through O. Thus the ideal points are incident with the ideal line. Under this interpretation, π' can be shown to be a model of the projective plane (see Exercise 5).

EXERCISES

1. Write out the proof of the dual of Axiom 1.

2. (a) Prove that there exist at least three points on every line of a projective plane. (Note: You cannot assume the existence of *any* points on a line.)
   (b) Extend your proof in part (a) to show that there exist at least four points on every line of a projective plane.

3. Find a model for the axiom system consisting of Axioms 1–3 that has exactly three points on every line. What is the total number of points in this model? The total number of lines? Does your model satisfy Axiom 4?

4. Show that Axiom 4 is independent of Axioms 1–3. (Note: Axiom 4 is known as *Fano's axiom.*)

5. Verify that π' satisfies Axioms 1–3. Which points in π' are points of intersection of lines that are parallel in the Euclidean plane π?

6. (a) List all possible quadrangles in model 3 of Section 1.3 that contain the points A and B as two of the four vertices. (b) Verify Axiom 4 for the four quadrangles in this model in which three of the vertices are A, B, and E.

4.3. Perspective Triangles

Although Axioms 1–4 describe the basic properties of our projective plane, we will require two more important properties, which are formalized in Axioms 5 and 6. The first of these properties concerns two relations between pairs of triangles. As the following definition indicates, one of these relations requires a correspondence between vertices and the other requires a correspondence between sides. As in the familiar case of congruent triangles in Euclidean geometry, the order in which the vertices of the triangles are named is used to indicate the correspondence.

Definition 4.4. Triangles $\triangle ABC$ and $\triangle A'B'C'$ are said to be *perspective from a point* if the three lines joining corresponding vertices, AA', BB', and CC', are concurrent. The triangles are said to be *perspective from a line* if the three points of intersection of corresponding sides, $AB \cdot A'B'$, $AC \cdot A'C'$, and $BC \cdot B'C'$, are collinear (see Fig. 4.5).

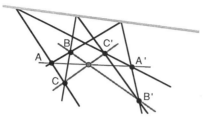

Figure 4.5

Axiom 5 (Desargues' Theorem). If two triangles are perspective from a point, they are perspective from a line.

This statement can be easily proved in projective geometry of 3-space (see Coxeter, 1987, *Projective Geometry*) and hence is frequently referred to as Desargues' Theorem, thus honoring the French mathematician who anticipated the development of projective geometry. However, in plane projective geometry, either this statement or an equivalent statement must be assumed as an axiom; since in some geometries that satisfy Axioms 1–4 this statement does not hold.

To ensure that our axiom system still satisfies the principle of duality, we must prove the dual of Axiom 5. In this case the dual is just the converse of the axiom.

Theorem 4.3 (Dual of Axiom 5). *If two triangles are perspective from a line, they are perspective from a point.*

Proof. Assume $\triangle ABC$ and $\triangle A'B'C'$ are perspective from a line, that is, $AB \cdot A'B' = P$, $B'C' \cdot BC = Q$, and $AC \cdot A'C' = R$ are collinear (see Fig. 4.6). It is sufficient to show that AA', BB', and CC' are concurrent. Let $O = AA' \cdot BB'$, and consider $\triangle RAA'$ and $\triangle QBB'$. Then P is on RQ since P, Q, and R are collinear and P is on AB and on $A'B'$ by definition of P. Thus, $\triangle RAA'$ and $\triangle QBB'$ are perspective from P, so by Axiom 5 they are perspective from a line; that is, $RA \cdot QB = C$, $RA' \cdot QB' = C'$, and $AA' \cdot BB' = O$ are collinear. Thus AA', BB', and CC' are concurrent. □

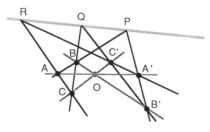

Figure 4.6

The importance of this axiom and its implications cannot be overstated. It provides a means of proving that three points are collinear and the preceding proof offers a typical example of this use. It will also be used to show that the fourth point of a set known as a harmonic set is unique.

EXERCISES

1. Construct two triangles that are perspective from a point. From which line are they perspective?

2. Construct two triangles that are perspective from a line. From which point are they perspective?

3. (a) Is the Desargues' configuration given in Section 1.5 a projective plane? Why? (b) Show that in this configuration, $\triangle BEH$ and $\triangle ADI$ are perspective from a point and from a line.

4. If the vertices of $\triangle PQR$ lie, respectively, on the sides of $\triangle ABC$ so that AP, BQ, and CR are concurrent, and if $AB \cdot PQ = U$, $AC \cdot PR = V$, and $BC \cdot QR = W$; show that U, V, and W are collinear.

4.4. Harmonic Sets

This section introduces special sets of four collinear points (and dual sets of four concurrent lines) that are defined entirely in terms of a construction involving points and lines. In Section 4.5, we shall see that point and line constructions can be used to define correspondences between two sets of collinear points, two sets of concurrent lines, and a set of collinear points and a set of concurrent lines; in Section 4.6, point and line constructions are used to define conics.

Definition 4.5. Four collinear points, A, B, C, D are said to form *harmonic set* $H(AB, CD)$ if there is a complete quadrangle in which two opposite sides pass through A, two other opposite sides pass through B, while the remaining two sides pass through C and D, respectively. C is called the *harmonic conjugate* of D (or D is the harmonic conjugate of C) with respect to A and B.

Note that A and B are diagonal points of the quadrangle and are named first. Also note that the points of the first pair in the harmonic set are distinguished from the points of the second pair but there is no distinction made between points of the first pair or points of the second pair; that is,

$$H(AB, CD) \Leftrightarrow H(BA, CD) \Leftrightarrow H(AB, DC) \Leftrightarrow H(BA, DC).$$

Using this definition, given any three distinct collinear points, A, B, C; then D, the harmonic conjugate of C with respect to A and B, can be constructed as follows.

Construction of the Fourth Point of a Harmonic Set

Let E be an arbitrary point not on AB and let m be a line through B that is distinct from AB and not incident with E. Let $m \cdot AE = F$, $m \cdot CE = G$, and $AG \cdot EB = H$. As you can verify, the points E, F, G, and H form a complete quadrangle with two opposite sides through A, two opposite sides through B, and one of the remaining sides through C (see Fig. 4.7). Therefore, $D = FH \cdot AB$.

Using Axiom 4, we can verify that D is distinct from A, B, and C (see Exercise 4), thus demonstrating again that each line of our projective plane contains at least four points.

Both the definition and the preceding construction for finding, D, the harmonic conjugate of C with respect to the points A and B, may make the point D appear somewhat arbitrary. However, the following theorem shows that if we begin with three given points A, B, and C, any construction that satisfies Definition 4.5 will give the same point D; that is, D is uniquely determined.

Theorem 4.4. *If A, B, and C are three distinct, collinear points, then D, the harmonic conjugate of C with respect to A and B, is unique.*

Proof. Let $EFGH$ be a quadrangle used to find the point D. Assume a second quadrangle $E'F'G'H'$ is also constructed so that $E'H' \cdot F'G' = B$, $E'F' \cdot G'H' = A$, and $E'G' \cdot AB = C$, and let $D^* = F'H' \cdot AB$ (see Fig. 4.8). It suffices to show that $D^* = D$. To do this, Axiom 5 and its dual are employed.

Note that $\triangle EFG$ and $\triangle E'F'G'$ are perspective from line AB. So by Theorem 4.3, they are perspective from a point; i.e., EE', FF', and GG' are concurrent. Similarly $\triangle EGH$ and $\triangle E'G'H'$ are perspective from AB and hence EE', GG', and HH' are concurrent. Thus the four lines EE', FF', GG', and HH' are all concurrent. So $\triangle FHG$ and $\triangle F'H'G'$ are perspective from a point, and by Axiom 5 it follows that they are perspective from a line. So $FH \cdot F'H'$, $FG \cdot F'G' = B$, and $HG \cdot H'G' = A$ are collinear. But $FH \cdot AB = D$, $F'H' \cdot AB = D^*$. Thus $D = D^*$.   □

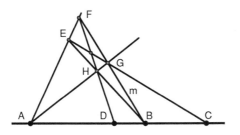

Figure 4.7

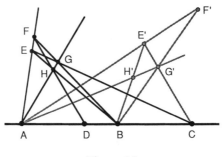

Figure 4.8

In addition to the possible order changes within the first and last pairs of points of a harmonic set, the following theorem indicates that the pairs themselves may be interchanged.

Theorem 4.5. $H(AB, CD) \Leftrightarrow H(CD, AB)$.

Proof. We assume $H(AB, CD)$ and show $H(CD, AB)$. A similar proof can be used to verify the second half of the equivalence.

Since $H(AB, CD)$, there is a quadrangle $EFGH$ such that $A = EF \cdot GH$, $B = EH \cdot FG$, $C = EG \cdot n$, and $D = FH \cdot n$ where $n = AB$. Now let $S = DG \cdot FC$ and $T = GE \cdot FH$, and consider quadrangle $TGSF$ (see Fig. 4.9). Note the two lines $SF = FC$ and $TG = GE$ are both incident with C. Also $GS = DG$ and $TF = FH$ are both incident with D. Furthermore, line GF is incident with B. Thus it suffices to show that TS is incident with A. Note that $A = EF \cdot GH$. So consider $\triangle THE$ and $\triangle SGF$. If these triangles can be shown to be perspective from a point, it immediately follows that A is incident with TS and therefore that $H(CD, AB)$.

Since the intersections of corresponding sides of these triangles are $TE \cdot SF = GE \cdot FC = C$, $TH \cdot SG = FH \cdot DG = D$, and $HE \cdot GF = B$; these triangles are perspective from line n and therefore they are perspective from a point.

□

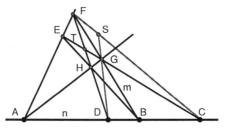

Figure 4.9

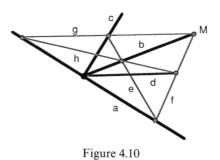

Figure 4.10

Corollary. $H(AB, CD) \Leftrightarrow H(AB, DC) \Leftrightarrow H(BA, CD) \Leftrightarrow H(BA, DC) \Leftrightarrow H(CD, AB) \Leftrightarrow$ $H(CD, BA) \Leftrightarrow H(DC, AB) \Leftrightarrow H(DC, BA)$.

As in previous sections, the dual of this definition of a harmonic set of points can be formulated.

Definition 4.6. Four concurrent lines, a, b, c, d, are said to form the *harmonic set* $H(ab, cd)$ if there is a complete quadrilateral in which two opposite vertices lie on a, two other opposite vertices lie on b, while the remaining two vertices lie on c and d, respectively (see Fig. 4.10 where lines e, f, g, and h form a quadrilateral yielding $H(ab, cd)$).

The construction of the fourth line of a harmonic set and the following theorems follow automatically by dualizing the previous results.

Theorem 4.6. *If lines $a, b,$ and c are concurrent, then d, the harmonic conjugate of c with respect to a and b, is unique.*

Theorem 4.7. $H(ab, cd) \Leftrightarrow H(cd, ab)$.

Eventually, we shall see that the harmonic property is an invariant under the transformations of projective geometry. In addition, the harmonic property can be used to coordinatize the projective plane, that is, using constructions involving only lines and points and without any notion of distance, a coordinate system can be constructed that assigns to each point in the projective plane an ordered pair of numbers. (For a detailed presentation of this process see Tuller.)

EXERCISES

1. Let points A, B, C be located as shown in Figs. 4.11 and 4.12. Construct the harmonic conjugate of C with respect to A and B: (a) in Fig. 4.11; and (b) in Fig. 4.12.

2. Let lines a, b, c be located as shown in Figs. 4.13 and 4.14. Construct the

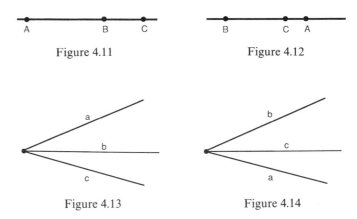

Figure 4.11   Figure 4.12

Figure 4.13   Figure 4.14

harmonic conjugate of c with respect to a and b: (a) in Fig. 4.13; and (b) in Fig. 4.14.

3. In Fig. 4.7, let $I = EG \cdot FH$. Show that I is an element of two nonequivalent harmonic sets in this figure. Be sure to specify the quadrangle involved in each.

4. Prove that the fourth point of a harmonic set is distinct from the three other points of the harmonic set, that is, if $H(AB, CD)$, prove that D is distinct from A, B, and C.

5. Suppose, in the Euclidean plane, that B is the midpoint of segment AC. Try to construct the harmonic conjugate of B with respect to A and C. What happens?

The following exercise is reprinted with permission from Coxeter (1987, *Projective Geometry*, p. 23).

6. Working in the Euclidean plane, draw a line-segment OC, take G two-thirds of the way along it, and E two-fifths of the way from G to C. (For instance, make the distances in centimeters $OG = 10$, $GE = 2$, $EC = 3$.) If the segment OC represents a stretched string, tuned to the note C, the same string stopped at E or G will play the other notes of the major triad. By drawing a suitable quadrangle, verify experimentally that $H(OE, CG)$. (Such phenomena explain our use of the word *harmonic*.)

4.5. Perspectivities and Projectivities

Transformations of the projective plane, known as collineations, are introduced analytically in Section 4.10. In that section we will see that, as the name suggests, these transformations preserve *collinearity*; that is, the images of collinear points are also collinear. Thus if we restrict our view to points on a particular line, we will be able to say that a collineation induces a mapping from this set of collinear points to another set of collinear points. As you may expect, we shall see that collineations preserve *concurrence* as well; that is, the images of concurrent lines will be concurrent lines. So collineations will also induce mappings from a set of concurrent lines to another set of concurrent lines. A second type of transformation, known as correlations, will induce mappings from collinear points to concurrent lines and vice versa.

In this section, we learn how to use point and line constructions to obtain, synthetically, correspondences, which we later show are exactly the correspondences given analytically by the induced mappings described earlier. Many of the terms and properties involved in these constructions reflect the artistic origins of projective geometry.

In order to facilitate our description of these constructions we will begin by adopting the following dual definitions.

Definition 4.7. The set of lines through a point P is called a *pencil of lines with center P* (see Fig. 4.15); the set of all points on a line p is called a *pencil of points with axis p* (see Fig. 4.16).

With these definitions, the mappings mentioned previously can be formally defined in terms of mappings between pencils. The most elementary of these mappings are known as perspectivities.

Figure 4.15   Figure 4.16

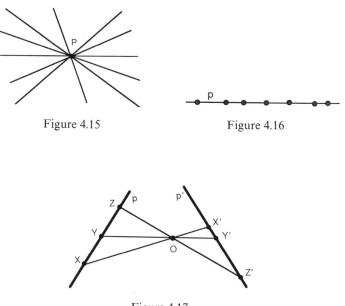

Figure 4.17

Definition 4.8a. A one-to-one mapping between two pencils of points with axes p and p' is called a *perspectivity* if each line joining the point X on p with the corresponding point X' on p' is incident with a fixed point O. O is called the *center of the perspectivity*. Such a perspectivity is denoted $X \overset{o}{\wedge} x'$ (see Fig. 4.17).

Definition 4.8b. A one-to-one mapping between two pencils of lines with centers P and P' is called a *perspectivity* if each point of intersection of the corresponding lines x on P and x' on P' lies on a fixed line o. o is called the *axis of the perspectivity*. Such a perspectivity is denoted $x \overset{o}{\overline{\wedge}} x'$ (see Fig. 4.18).

Definition 4.8c. A one-to-one mapping between a pencil of points with axis p and a pencil of lines with center P is called a *perspectivity* if each point X on p is incident with the corresponding line x on P. Such a perspectivity is denoted $X \overline{\wedge} x$ or $x \overline{\wedge} X$ (see Fig. 4.19).

Note that in Definition 4.8c, the first pencil can be a pencil of points, and the image pencil a pencil of lines; or the first pencil can be a pencil of lines, and the image pencil a pencil of points.

In each of the three definitions, the pencils are said to be *perspectively related*. If the perspectively related pencils are of the same kind, we can show (see Exercise 1) that the perspectivity is uniquely determined by two pairs of corresponding elements (provided no element of the two pairs is on both pencils). In other words, once two pairs of corresponding elements are

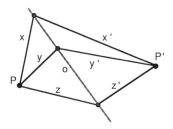

Figure 4.18

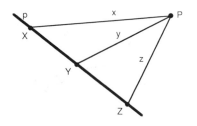

Figure 4.19

specified, the image of any third element of the first pencil is uniquely determined.

Since perspectivities are one-to-one mappings, their inverses exist and are clearly again perspectivities. Also a finite product of perspectivities, that is, a finite number of perspectivities used in succession, produces another mapping known as a *projectivity*. It is to these mappings that our final axiom refers.

Definition 4.9. A one-to-one mapping between the elements of two pencils is called a *projectivity* if it consists of a finite product of perspectivities.

Figures 4.20–4.22 show projectivities between pencils of points, pencils of lines, and a pencil of lines and a pencil of points, respectively. Notice that the notation for projectivities uses the *unadorned* symbol " ∧ ." When a projectivity exists between two pencils, the pencils are said to be *projectively related*. If a projectivity maps either a pencil of points or a pencil of lines onto itself, it is called a *projectivity on the pencil*. Axiom 6, which is self-dual, describes an important property of projectivities on pencils.

Axiom 6. If a projectivity on a pencil leaves three elements of the pencil invariant, it leaves every element of the pencil invariant.

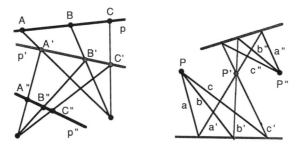

Figure 4.20. $ABC \wedge A''B''C''$. Figure 4.21. $abc \wedge a''b''c''$.

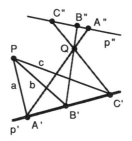

Figure 4.22. $abc \wedge A''B''C''$.

Thus the projectivity on a pencil that keeps three elements invariant is necessarily the identity mapping. Other observations that should be made at this point include: (1) a projectivity does not have a center or an axis unless it consists of just one perspectivity; and (2) the inverse of a projectivity and the product of two projectivities are again projectivities.

Whereas a perspectivity between two pencils is uniquely determined by two pairs of corresponding elements of the pencils, the existence of a projectivity between two pencils that maps any three elements of the first pencil to three corresponding elements of the second pencil can be demonstrated by construction. This construction is demonstrated for two distinct pencils of points.

Construction of a Projectivity Between Pencils of Points

Let A, B, C be elements of the pencil with axis p and A', B', C' corresponding elements of the pencil with axis p' ($p \neq p'$). Construct line AA' and choose a point $P \neq A'$ on this line. Let $m \neq p'$ be an arbitrary line through A'. Let $B_1 = BP \cdot m$, $C_1 = CP \cdot m$. Thus, $ABC \overset{P}{\wedge} A'B_1C_1$. Now let $Q = B_1B' \cdot C_1C'$. Then $A'B_1C_1 \overset{Q}{\wedge} A'B'C'$ and therefore $ABC \wedge A'B'C'$ (see Fig. 4.23).

Note that the preceding construction requires only two perspectivities, but the construction of these perspectivities is not unique.

The existence of a projectivity between pencils of lines that maps any three lines of the first pencil to three corresponding lines of the second pencil follows by duality. The existence of a projectivity that maps three concurrent lines to three corresponding collinear points can also be easily demonstrated (see Exercise 4).

Thus any three elements of one pencil can be projectively related to three arbitrary elements of a second pencil and the correspondence defined by the projectivity constructed from these three pairs, can be extended to pair all the remaining elements of the two pencils. However, since the construction involved is not uniquely specified, it is not immediately apparent that the

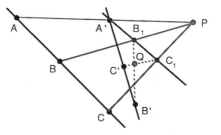

Figure 4.23

images of any fourth element of the first pencil determined by different constructions always turn out to be the same. The remarkable result that says this does indeed happen is known as the *fundamental theorem of projective geometry*.

Theorem 4.8 (Fundamental Theorem). *A projectivity between two pencils is uniquely determined by three pairs of corresponding elements.*

Proof. The existence of a projectivity has been demonstrated. The uniqueness follows from Axiom 6 as shown.

Case 1. Two pencils of points. Assume that A, B, C are elements of a pencil of points with axis p and that A', B', C' are the corresponding elements of a second pencil with axis p'. By the preceding result, there exists a projectivity T such that:

$$T: ABC \to A'B'C'.$$

If T is not unique, there exists another projectivity S, such that

$$S: ABC \to A'B'C'$$

Then under the projectivity ST^{-1}

$$A'B'C' \wedge ABC \wedge A'B'C'$$

or in other words ST^{-1} is a projectivity on p' keeping the three points A', B', and C' invariant. Therefore, by Axiom 6, $ST^{-1} = I$ or $S = T$.

Case 2. Two pencils of lines. The proof follows automatically by duality of case 1.

Case 3. A pencil of points and a pencil of lines. This case follows from cases 1 and 2 and the use of a perspectivity between a pencil of points and a pencil of lines that is also uniquely determined. ☐

The construction used to demonstrate the existence of a projectivity mapping three elements of one pencil to three corresponding elements of a second pencil also provides direct proofs of two corollaries to the fundamental theorem (see Exercise 8).

Corollary 1. *If in a projectivity between two distinct pencils an element corresponds to itself, then the projectivity is a perspectivity (i.e., the mapping requires only one perspectivity).*

Corollary 2. *A projectivity between two pencils can be expressed as the product of at most three perspectivities.*

Since projectivities are mappings induced by the general transformations of the projective plane, it is important to note that the harmonic relation remains invariant under projectivities.

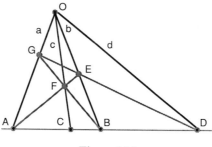

Figure 4.24

Theorem 4.9. *The harmonic relation is invariant under a projectivity. So, for example, if* $H(AB, CD)$ *and* $ABCD \wedge A'B'C'D'$, *then* $H(A'B', C'D')$.

Proof. Since the projective plane possesses duality, and any projectivity is a product of perspectivities, it is sufficient to show that $H(AB, CD)$ implies $H(ab, cd)$ where $ABCD \barwedge abcd$. Let $O = a \cdot b$ thus $a = OA, b = OB$, and so on. Since $H(AB, CD)$, there exists a quadrangle with one vertex at O, $OEFG$, such that A and B are diagonal points of the quadrangle, etc. Let $A = EF \cdot OG$, $B = OE \cdot GF$, $C = OF \cdot AB$, and $D = GE \cdot AB$ (see Fig. 4.24). Now consider quadrilateral GF, GE, AE, AB. Then $GF \cdot GE = G$, and $AE \cdot AB = A$ are on a; $GE \cdot AE = E$ and $GF \cdot AB = B$ are on b; $GE \cdot AB = D$ is on d, and $GF \cdot AE = F$ is on c. Thus $H(ab, cd)$.   □

As we have seen, three elements of one pencil can always be mapped to three elements of a second pencil via a projectivity, but a set of four elements of one pencil cannot in general be mapped to a set of four elements of a second pencil. If, however, both the first and second sets are harmonic sets, the desired projectivity will exist. This is formalized in the following theorem, which holds for both pencils of points and pencils of lines, even though the notation used is suggestive of pencils of points.

Theorem 4.10. *If four elements of one pencil,* A, B, C, D, *form a harmonic set,* $H(AB, CD)$, *and four elements of a second pencil,* A', B', C', D', *form a second harmonic set,* $H(A'B', C'D')$, *then there exists a projectivity mapping* A, B, C, D *to* A', B', C', D', *respectively.*

Proof. By Theorem 4.8, there is a projectivity such that $ABC \wedge A'B'C'$. Let D^* be the image of D under this projectivity. Then by Theorem 4.9, $H(A'B', C'D^*)$; but by Theorem 4.4, the harmonic conjugate of C' with respect to A' and D' is unique. Thus $D^* = D'$.   □

Before leaving the topic of projectivities, it is useful to note that there is a second, frequently more convenient, method for constructing the images under a projectivity between pencils of points. This method makes use of the following definition and theorem.

Definition 4.10. If A and A', B and B' are pairs of corresponding points, the *cross joins* of these pairs of points are the lines AB' and BA'.

Theorem 4.11. *A projectivity between two distinct pencils of points determines a unique line called the* axis of homology, *which contains the intersections of the cross joins of all pairs of corresponding points.*

Proof. Consider two distinct pencils of points with axes p and p'. Assume $ABC \wedge A'B'C'$, where $P = p \cdot p'$ is none of the six points. Clearly $A'A$, $A'B$, $A'C \barwedge ABC$ and $A'B'C' \barwedge AA'$, AB', AC'. Thus $A'A$, $A'B$, $A'C \wedge AA'$, AB', AC', so by Corollary 1, of the fundamental theorem $A'A$, $A'B$, $A'C \overset{h}{\barwedge} AA'$, AB', AC' for some axis h. So $A'B \cdot AB'$ and $A'C \cdot AC'$ are both on h (see Fig. 4.25).

To use h to find the image of another point D on p, proceed as follows. Construct $A'D$. Let $D_1 = A'D \cdot h$. Then $D' = AD_1 \cdot p'$.

To show that h is unique, it is necessary to show that h is independent of the choices for the centers of the pencils of lines (here A and A') and thus that the intersections of cross joins of all pairs of corresponding points are on h. To do this it is sufficient to find two points on h that are independent of these choices. Let $Q = h \cdot p'$ and $R = h \cdot p$. Using the technique described earlier to locate the image of R, let $R_1 = A'R \cdot h$. But $A'R \cdot h = R$. So $R_1 = R$. Then $R' = AR \cdot p' = p \cdot p' = P$; that is, the image of R is P. Likewise, the image of P can be shown to be Q. But the image and preimage of P are uniquely determined by Theorem 4.8. (Note that $Q \neq R$ since this projectivity is not a perspectivity.) Thus $h = QR$ is uniquely determined.   □

The proof of Theorem 4.11 contains a description of the method used to construct h and find the image of any arbitrary point. Clearly the cross joins of two pairs of lines can be defined by dualizing Definition 4.10 and the dual of Theorem 4.11 can be used to construct the images of lines under projectivities between pencils of lines using a *center of homology*.

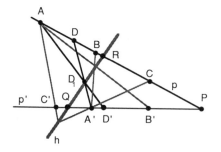

Figure 4.25

1. Prove that a perspectivity between two pencils of the same kind is uniquely determined by two pairs of corresponding elements (provided no element of the two pairs is on both pencils).

2. Given a perspectivity between two distinct pencils of points with axes p and p', verify each of the following: (a) the center of the perspectivity is not incident with either p or p'; and (b) the point $P = p \cdot p'$ maps to itself under this perspectivity.

3. Demonstrate the existence of a projectivity mapping concurrent lines a, b, c in a pencil with center P to concurrent lines a', b', c' in a pencil with center P' (assume $P \neq P'$).

4. Demonstrate the existence of a projectivity mapping concurrent lines a, b, c to collinear points A, B, C.

5. Let a, b, c be three concurrent lines and P, Q two points not on any of them. Let $A_1, A_2, \ldots$ and $B_1, B_2, \ldots$ be points on a and b, respectively, such that $A_i P \cdot B_i Q = C_i$ where C_i is on line c. Show that $A_i \wedge B_i$.

6. Given four distinct collinear points A, B, C, D, construct the following projectivities: (a) $ABC \wedge ABD$; (b) $ABC \wedge ACD$; (c) $ABC \wedge BAD$; (d) $ABC \wedge ACB$. Find the image of D under part (d).

7. What is the minimum number of perspectivities required for each part in Exercise 5?

8. Prove the corollaries to Theorem 4.8.

9. Use the dual of Theorem 4.11 to find the center of homology determined by two projectively, but nonperspectively related pencils of lines. Demonstrate the construction of an image line.

4.6. Conics in the Projective Plane

So far in Chapter 4, we have studied projective figures determined by sets of n points, where no three of these points are collinear. For $n = 3$ we considered figures known as triangles, and for $n = 4$ we considered figures known as quadrangles. We now consider figures that we shall eventually discover are uniquely determined by such sets when $n = 5$. These figures, known as *point conics* are defined in terms of projectivities.

Definition 4.11. A *point conic* is the set of points of intersection of corresponding lines of two projectively, but not perspectively, related pencils of lines with distinct centers (see Fig. 4.26).

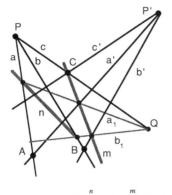

Figure 4.26. $abc \stackrel{n}{\barwedge} a_1b_1c \stackrel{m}{\barwedge} a'b'c'$.

It is not obvious that the point conics just defined are in any way related to the familiar conics of Euclidean geometry. In this section, we demonstrate that point conics are determined uniquely by five points, no three collinear (Theorem 4.14), but the connection between the point conics of projective geometry and Euclidean conics does not become apparent until much later. However, the following definition of a *tangent* does resemble the familiar Euclidean definition.

Definition 4.12. A *tangent to a point conic* is a line that has exactly one point in common with the point conic.

Both the definition of a point conic and the definition of a tangent can be dualized to define other concepts in projective geometry. The figure described by the dual of Definition 4.11 is known as a *line conic* and the point described by the dual of Definition 4.12 is known as a *point of contact*. With these definitions, each of the theorems describing properties of point conics developed in this section can be dualized to describe corresponding properties of line conics.

Definition 4.13. A *line conic* is the set of lines joining corresponding points of two projectively, but not perspectively, related pencils of points with distinct axes.

Definition 4.14. A *point of contact of a line conic* is a point that lies on exactly one line of the line conic.

As Definition 4.11 indicates, a point conic is determined by a projectivity between two pencils of lines, and as stated by the fundamental theorem, these mappings are uniquely determined when three pairs of corresponding lines are specified. Thus, given pencils of lines with centers at P and P' ($P \neq P'$), we can arbitrarily pick three lines a, b, and c incident with P and three corresponding lines a', b', and c' incident with P'. Provided this correspondence does not yield a perspectivity, we can immediately locate three points of the point conic determined by this projectivity, namely, $a \cdot a'$, $b \cdot b'$, and $c \cdot c'$. (Note that different choices of lines and/or their corresponding lines will yield different point conics.) The following theorem shows that there are two more easily obtainable points of this point conic. However, other points in addition to these five must be located either by a construction of the projectivity or by other constructions described later in this section.

Theorem 4.12. *The centers of the pencils of lines in the projectivity defining a point conic are points of the point conic.*

Proof. Let P and P' be the centers of the pencils. Let $m = PP'$, and consider m as a line in the pencil with center P (see Fig. 4.27). Then there is a corresponding line m' in the pencil with center P'. Note that $m \neq m'$ since the projectivity is not a perspectivity. So $m \cdot m' = P'$ is a point of the point conic. Similarly, by considering m as a line in the pencil with center P', and finding the corresponding line, P can be shown to be a point of the point conic.   □

As a result of this theorem, any five points P_1, P_2, P_3, P_4, P_5 (no three collinear) can be used to determine a conic as follows.

Choose two of the points, say P_1 and P_2, as centers of pencils and construct lines P_1P_3, P_1P_4, P_1P_5 and P_2P_3, P_2P_4, P_2P_5. Then the projectivity P_1P_3, P_1P_4, $P_1P_5 \wedge P_2P_3$, P_2P_4, P_2P_5 defines a point conic containing the five points.

The point conic obtained by this construction clearly contains the original set of five points. To show that such a set of five points *uniquely* determines a point conic, we will use a projective figure consisting of six points (here, however, we do *not* require the condition that no three of the points are collinear).

Definition 4.15. A *hexagon* is a set of six distinct points called *vertices*, say, P_1, P_2, P_3, P_4, P_5, P_6, and the six lines P_1P_2, P_2P_3, P_3P_4, P_4P_5, P_5P_6, and P_6P_1

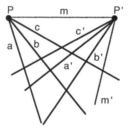

Figure 4.27

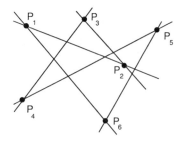

Figure 4.28. Hexagon $P_1P_2P_3P_4P_5P_6$.

(see Fig. 4.28). These lines are called the *sides* of the hexagon $P_1P_2P_3P_4P_5P_6$. Points P_1 and P_4, P_2 and P_5, P_3 and P_6 are pairs of *opposite vertices* and lines P_1P_2 and P_4P_5, P_2P_3 and P_5P_6, P_3P_4 and P_6P_1 are pairs of *opposite sides*. The three points of intersection of opposite sides are *diagonal points*.

It is important to observe that a given set of six points does not determine a unique hexagon, since a hexagon is determined by the order in which its vertices are named. In fact, a given set of six points can determine $6!/12 = 60$ different hexagons (see Exercise 3). Thus, in Theorem 4.13 it is important to notice that P and P', the centers of the pencils used to define the point conic, are used as the first and third vertices of the hexagon, respectively.

Theorem 4.13. *If A, B, C, D are four points on a point conic defined by projectively related pencils with centers P and P', then the diagonal points of hexagon $PBP'ACD$ are collinear, and conversely, if the diagonal points of hexagon $PBP'ACD$ are collinear, then A, B, C, D are points of the point conic determined by the projectively related pencils with centers P and P'.*

Proof. (a) The diagonal points for hexagon $PBP'ACD$ are $PB \cdot AC = J$, $BP' \cdot CD = L$, and $P'A \cdot DP = K$. Let $AC \cdot PD = M$ and $AP' \cdot DC = N$ (see Fig. 4.29). By using these and the definition of a point conic, we obtain the following projectivities: $AJCM \wedge PA$, PB, PC, $PD \wedge P'A$, $P'B$, $P'C$, $P'D \wedge NLCD$, or $AJCM \wedge NLCD$. But since $C \wedge C$, this projectivity is a perspectivity. And since $AN \cdot MD = AP' \cdot PD = K$, the center of the perspectivity is K. Thus J, L, and K are collinear.

(b) The proof of the converse is merely a reverse argument of the previous proof.   □

Using this result, we can now show that a set of five points, no three collinear, uniquely determines a point conic. This means that the constructions determined by using different pairs of the five points as centers of the projectively related pencils all yield the same set of points.

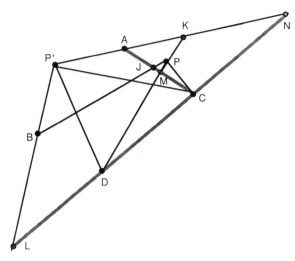

Figure 4.29

Theorem 4.14. *A point conic is uniquely determined by five distinct points, no three of which are collinear.*

Proof. Let P_1, P_2, P_3, P_4, P_5 be five points, no three collinear. Then there exists a point conic determined by the pencils with centers P_1 and P_2 and the projectivity $P_1P_3, P_1P_4, P_1P_5 \wedge P_2P_3, P_2P_4, P_2P_5$, which contains these five points. Let D be any sixth point on this point conic. To show that the conic is uniquely determined, that is, that the same set of points is determined when points other than P_1 and P_2 are used as the centers of the pencils, it is sufficient to show that D is on the point conic defined by pencils with centers at any two of the other points. Consider hexagon $P_1P_4P_2P_3P_5D$. By Theorem 4.13, the diagonal points $P_1P_4 \cdot P_3P_5$, $P_4P_2 \cdot P_5D$, $P_2P_3 \cdot DP_1$ are collinear. But this hexagon is the same as hexagon $P_4P_2P_3P_5DP_1$ and thus by the second part of Theorem 4.13, D is on the point conic determined by pencils with centers P_3 and P_4. By similarly renaming this hexagon or using other hexagons with P_1 and P_2 as the first and third vertices, it can be shown that D is on the point conic determined by pencils with centers at any two of the points P_1, P_2, P_3, P_4, P_5. □

This theorem has several interesting corollaries. The first of these is known by the intriguing title *Pascal's mystic hexagram theorem* and was proved by Pascal in 1640, when he was 17. The dual of this corollary was not proved until 1806 when Brianchon developed its proof.

Corollary 1 (Pascal's Theorem). *If a hexagon is inscribed in a point conic (i.e., the vertices of the hexagon are points of the point conic), its diagonal points are collinear (see Fig. 4.30).*

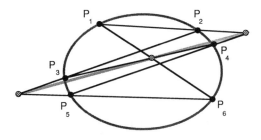

Figure 4.30

By considering hexagon $P'_1 P_1 P_2 P_3 P_4 P_5$ and letting point P'_1 approach P_1 so that line $P'_1 P_1$ becomes the tangent at P_1, we can verify a second corollary that gives an efficient method for constructing tangents to a point conic. Applying a similar process to two hexagons, namely, $P_1 P_2 P'_2 P_4 P_3 P'_3$ and $P_1 P'_1 P_2 P_4 P'_4 P_3$, yields a third corollary (see Exercise 8).

Corollary 2. *If the five points* P_1, P_2, P_3, P_4, P_5 *are points of a point conic, then the three points* $P_1 P_2 \cdot P_4 P_5$, $P_2 P_3 \cdot P_5 P_1$, $P_3 P_4 \cdot tangent(at\ P_1)$, *are collinear (see Fig. 4.31).*

Corollary 3. *If* P_1, P_2, P_3, P_4 *are four points of a point conic, then the four points* $P_1 P_2 \cdot P_3 P_4$, $P_1 P_3 \cdot P_2 P_4$, $\tan P_2 \cdot \tan P_3$, *and* $\tan P_1 \cdot \tan P_4$ *are collinear (see Fig. 4.32).*

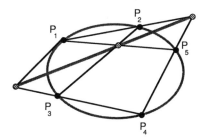

Figure 4.31

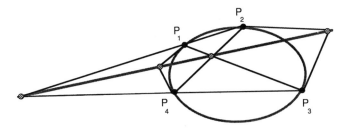

Figure 4.32

The construction of additional points of a point conic by setting up a construction for the projectivity involved is a fairly tedious procedure. However, the process can be simplified somewhat by using a center of homology as described by the dual of Theorem 4.11. A third method of constructing additional points uses Pascal's theorem as follows.

Construction of Points of a Point Conic Using Pascal's Theorem

Let A, B, C, D, E be five points of a point conic. Then any additional point, F, on the point conic can be considered as the sixth point of inscribed hexagon, $ABCDEF$. Since the diagonal points $P = AB \cdot DE$, $Q = BC \cdot EF$, and $CD \cdot FA$ will be collinear, choose a line m through E (this will be the line EF). Construct points P and Q. Then $R = CD \cdot PQ$ and $F = RA \cdot m$ (see Fig. 4.33). To locate other points on the point conic, merely choose other lines through E.

Even though the second corollary of Theorem 4.14 describes an easy method for constructing a tangent at a specific point, it does not given any insight into how tangent lines to a point conic are related to the projectivity defining the point conic. The proof of the following theorem not only demonstrates this relation, but also leads to a corollary we will use in Section 4.11 to find an equation for a point conic.

Theorem 4.15. *For any point A of a point conic, there is exactly one line tangent to the conic at A. (This tangent is the line corresponding to line AB considered as a line of the pencil through B when the conic is defined by projectively related pencils with centers A and B.)*

Proof. Let B, C, D, E be four more points of the point conic. Then the point conic can be defined by projectively related pencils with centers A and B. Let h be the line in the pencil with center A that corresponds to line AB considered as a line in the pencil with center B. Clearly h contains the point A of the point

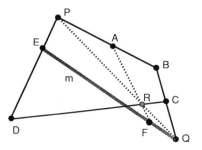

Figure 4.33

conic, and $h \cdot AB = A$. Assume h contains a second distinct point of the point conic, say, X.

Case 1. X is on AB. Then $h = AB$ and h corresponds to itself under the projectivity, and hence by Corollary 1 of Theorem 4.8, the projectivity is a perspectivity contradicting the definition of a point conic.

Case 2. X is not on AB. Then $h = AX$ corresponds to line AB and to line BX, which is distinct from AB. But this contradicts the one-to-one property of projectivities. Thus h contains exactly one point of the point conic and is, therefore, a tangent.

To show that there is no other tangent to the point conic at A, assume that a second line h' is also a tangent at A. Since $h' I A$, there is a line m in the pencil with center B that corresponds to h'. Then $m \cdot h'$ is a point of the point conic. But since h' is a tangent at A, it contains only one point of the conic, namely, A. Thus $m \cdot h' = A$ so $m = AB$, and hence $h' = h$.   □

Corollary. *A point conic is uniquely determined by three distinct noncollinear points and the tangents at two of them.*

By definition, tangents are the lines that intersect a conic exactly once. Other lines may or may not intersect the conic, but as the next theorem shows, a line can never intersect a conic *more* than twice. This result will be used in Section 4.11 when we use an analytic approach to study further properties of point conics.

Theorem 4.16. *A line intersects a point conic in at most two points.*

Proof. Assume line n intersects a point conic in three distinct points Q, R, and S. Let P and P' be two other points of the conic and consider the pencils with centers P and P'. Then, as shown previously, the conic can be defined in terms of a projectivity between these pencils where Q, R, and S are points of intersection of the corresponding pairs of lines PQ and $P'Q$, PR and $P'R$, PS and $P'S$. Under this projectivity, PQ, PR, $PS \wedge P'Q$, $P'R$, $P'S$. However, the three points Q, R, and S all lie on n, so in fact PQ, PR, $PS \overset{n}{\wedge} P'Q$, $P'R$, $P'S$. It follows by the fundamental theorem that the projectivity is a perspectivity, contradicting the definition of a point conic.   □

EXERCISES

1. Given five points, no three collinear, construct two more points of the point conic they determine and a tangent at one of the original five points using each of the following methods: (a) construction of the projectivity as a product of two perspectivities; and (b) using the center of homology.

2. Dualize Exercise 1, and perform the construction.

3. Explain why six points determine 60 different hexagons.

4. Prove: If alternate vertices of a hexagon lie on two lines (i.e., in hexagon $P_1P_2P_3P_4P_5P_6$, P_1, P_3, and P_5 are collinear as are P_2, P_4, and P_6) then the diagonal points are collinear. (This is known as *Pappus' Theorem* and dates from the 3rd century) [*Hint*: Find a projectivity between the two lines for which the points of intersection of the cross joins are the diagonal points of the hexagon.]

5. Given five points, no three collinear, construct two more points of the point conic they determine using Pascal's theorem.

6. Dualize Exercise 5, and perform the construction.

7. Prove that the tangent to a point conic at A is the line joining A to the center of homology determined by a projectivity between two pencils defining the conic where A is the center of one of the pencils.

8. Prove Corollary 3 of Theorem 4.14.

9. Show that omitting the phrase "but not perspectively" from Definition 4.11 would allow inclusion of sets of two lines (i.e., the points on these lines) as point conics. Which two lines would they be?

4.7. An Analytic Model for the Projective Plane

Until now we have considered plane projective geometry from a strictly synthetic point of view. We now change our point of view and adopt the approach suggested by Klein's definition of geometry; that is, we begin exploring the invariants of the projective plane under a group of transformations. To obtain matrix representations of these projective transformations, we need an analytic model of the projective plane. Since our goal is to look at the real projective plane, we consider an analytic model of the projective plane similar to the model of the Euclidean plane. Our matrix representations then resemble the matrices that we used for isometries, similarities, and affinities, so we are able to use techniques similar to those used in Chapter 3. Thus this approach enables us to both explore additional properties of the real projective plane and view projective geometry as the next logical step in the progression from Euclidean to similarity to affine geometry.

Our analytic model for the projective plane uses the nonzero equivalence classes determined by the relation on R^3 defined in Section 3.2 not only as lines but also as points. [Recall that $(a_1, a_2, a_3) \sim (b_1, b_2, b_3)$ if there is a nonzero real number k such that $(a_1, a_2, a_3) = k(b_1, b_2, b_3)$.] As in Chapter 3, ordered triples used to represent points are denoted with parentheses as (x_1, x_2, x_3) while ordered triples used to represent lines are denoted with square brackets as $[u_1, u_2, u_3]$. It is important to observe that the additional restrictions required for the interpretations of point and line in the Euclidean model are no longer required.

Analytic Model

Undefined Term	Interpretation
Point	A nonzero equivalence class of ordered triples of real numbers; any element (x_1, x_2, x_3) of the equivalence class will be called *homogeneous coordinates* of the point
Line	A nonzero equivalence class of ordered triples of real numbers; any element $[u_1, u_2, u_3]$ of the equivalence class will be called *homogeneous coordinates* of the line
Incidence	The line u is said to be incident with the point X if the dot product $u \cdot X = 0$, or in matrix notation, $$[u_1, u_2, u_3] \begin{bmatrix} x_1 \\ x_2 \\ x_3 \end{bmatrix} = 0$$

To show that this set of interpretations is a model of our projective plane, it is necessary to verify that it satisfies Axioms 1–6. The verification of Axioms 1–3 is left as an exercise (see Exercise 2). We verify Axiom 5 at the end of this section but postpone the verification of Axioms 4 and 6 to Sections 4.10 and 4.8, respectively.

We can visualize this analytic model in terms of the extended Euclidean plane π' introduced in Section 4.2. There we saw how to obtain π' from π (a Euclidean plane parallel to, but distinct from, the plane $x_3 = 0$) by extending the following correspondence between the set of points and lines in π and the set of lines and planes through the origin in E^3:

1. A point P in π corresponds to the line through the origin that intersects π at P.
2. A line l in π corresponds to the plane through the origin that intersects π along l.

To show that this correspondence yields homogeneous coordinates for the points and lines of π' we need to recall several analytic geometry facts about E^3 (Euclidean 3-space). In particular the following observations are useful:

1. Any line through the origin can be represented in vector notation as $(x_1, x_2, x_3) = t(s_1, s_2, s_3)$, where $\mathbf{x} = (x_1, x_2, x_3)$ is a vector from the origin to an arbitrary point X on the line and $\mathbf{s} = (s_1, s_2, s_3)$ is a direction vector for the line. (Note that any nonzero scalar multiple of a direction vector $\mathbf{s}$ is also a direction vector for the same line.)
2. Any plane through the origin can be represented by an equation of the form $\mathbf{n} \cdot \mathbf{x} = n_1 x_1 + n_2 x_2 + n_3 x_3 = 0$, where $\mathbf{x} = (x_1, x_2, x_3)$ is a vector from the

origin to an arbitrary point X on the plane and $\mathbf{n} = (n_1, n_2, n_3)$ is a vector normal (i.e., perpendicular) to the plane. (Note that any nonzero scalar multiple of a normal vector $\mathbf{n}$ is also a normal vector for the same plane.)
3. Thus a line through the origin with direction vector $\mathbf{s}$ will lie in a plane through the origin with normal vector $\mathbf{n}$ iff $\mathbf{n} \cdot \mathbf{s} = 0$.

We can then identify each point P in π with a nonzero equivalence class of R^3, namely, the set of all possible direction vectors for the line through the origin that intersects π at P. Likewise, we can identify each line l in π with a nonzero equivalence class of R^3, namely, the set of all possible normal vectors for the plane through the origin that intersects π at l. In this way, elements of the equivalence classes become the homogeneous coordinates of the points and lines in π.

To complete the process, we need to find homogeneous coordinates for the ideal points and line added to π to obtain π'. We can do this by identifying the ideal points with the nonzero equivalence classes that give direction vectors for lines through the origin that do not intersect π, and by identifying the ideal line added to π with the equivalence class of normal vectors for the plane $x_3 = 0$. It is interesting to note the form of the homogeneous coordinates for these ideal points and the ideal line (see Exercise 3).

Using this identification, it should become apparent that points in π' are collinear iff the corresponding lines through the origin in E^3 are coplanar, but a result from linear algebra says these lines in E^3 are coplanar iff their direction vectors are linearly dependent. Likewise, lines in π' are concurrent iff the corresponding planes through the origin in E^3 intersect along a common line, but this happens iff their normal vectors are linearly dependent. These observations anticipate the following results, which give algebraic conditions for the collinearity of points and concurrence of lines. (The proofs of these results are nearly identical to those used in Chapter 3.)

Theorem 4.17. *Three points X, Y, Z are collinear iff the determinant*

$$\begin{vmatrix} x_1 & y_1 & z_1 \\ x_2 & y_2 & z_2 \\ x_3 & y_3 & z_3 \end{vmatrix} = 0$$

Corollary. *The equation of the line PQ can be written*

$$\begin{vmatrix} x_1 & p_1 & q_1 \\ x_2 & p_2 & q_2 \\ x_3 & p_3 & q_3 \end{vmatrix} = 0$$

The dual statements give algebraic methods for determining when three lines are concurrent, and for finding the equation of a point determined by two lines. Here, however, the coordinates of the lines are used as rows rather than columns.

Theorem 4.18. *Three lines u, v, w are concurrent iff the determinant*

$$\begin{vmatrix} u_1 & u_2 & u_3 \\ v_1 & v_2 & v_3 \\ w_1 & w_2 & w_3 \end{vmatrix} = 0$$

Corollary. *The equation of the point p·q can be written*

$$\begin{vmatrix} u_1 & u_2 & u_3 \\ p_1 & p_2 & p_3 \\ q_1 & q_2 & q_3 \end{vmatrix} = 0$$

EXAMPLE 4.1. Find the equation of the point of intersection of lines $p[-2, 5, 7]$ and $q[3, 1, 2]$.

Using the corollary to Theorem 4.18, we can find the equation of the point by setting the following determinant equal to 0:

$$\begin{vmatrix} u_1 & u_2 & u_3 \\ -2 & 5 & 7 \\ 3 & 1 & 2 \end{vmatrix} = 0$$

Expanding this determinant results in the equation $3u_1 + 25u_2 - 17u_3 = 0$, which is the equation of a point. Note that the coordinates of this point are $(3, 25, -17)$.

In this model we can now show that projectivities between pencils can be represented via 2×2 matrices. (The analytic form of the transformations of the entire projective plane, which induce projectivities, will require 3×3 matrices and be developed later.) This matrix representation of projectivities requires that points and lines be assigned ordered pairs of real numbers rather than ordered triples. This is done by picking *base elements* for a pencil and making use of the following theorem.

Theorem 4.19. *If $P(p_1, p_2, p_3)$ and $Q(q_1, q_2, q_3)$ are two distinct points, any point R of the line PQ has homogeneous coordinates (r_1, r_2, r_3) where $r_i = \lambda_1 p_i + \lambda_2 q_i$, $i = 1, 2, 3$ and λ_1, λ_2 are real but not both 0, and conversely any point R with homogeneous coordinates of this form is on line PQ.*

Proof. (a) Assume R has homogeneous coordinates $(\lambda_1 p_1 + \lambda_2 q_1, \lambda_1 p_2 + \lambda_2 q_2, \lambda_1 p_3 + \lambda_2 q_3)$, then

$$\begin{vmatrix} r_1 & p_1 & q_1 \\ r_2 & p_2 & q_2 \\ r_3 & p_3 & q_3 \end{vmatrix} = \begin{vmatrix} \lambda_1 p_1 + \lambda_2 q_1 & p_1 & q_1 \\ \lambda_1 p_2 + \lambda_2 q_2 & p_2 & q_2 \\ \lambda_1 p_3 + \lambda_2 q_3 & p_3 & q_3 \end{vmatrix} = 0$$

So by Theorem 4.17, the points P, Q, and R are collinear.

(b) If R is on PQ then $|PQR| = 0$, or in other words, the vectors corresponding to these three points are linearly dependent. Thus there exist real numbers $\lambda_1, \lambda_2, \lambda_3$, not all zero, such that $\lambda_1 P + \lambda_2 Q + \lambda_3 R = 0$. Note that $\lambda_3 \neq 0$ since P and Q are distinct points; therefore, assume $\lambda_3 = -1$, thus $\lambda_1 P + \lambda_2 Q = R$.

Definition 4.16. The points P and Q used in Theorem 4.19 are called *base points*, while λ_1 and λ_2 are called *homogeneous parameters* of R with respect to P and Q.

Clearly the homogeneous parameters of the base points P and Q are $(1, 0)$ and $(0, 1)$, respectively. In general, the homogeneous parameters of a point depend on the base points chosen and on their homogeneous coordinates. So specific homogeneous coordinates for the base points must be used (see Exercise 9). Even so, there is not a unique set of homogeneous parameters for each point, since (λ_1, λ_2) and $(k\lambda_1, k\lambda_2)$ represent the same point $(k \neq 0)$; but the ratio $\lambda = \lambda_1/\lambda_2$ is unique. This ratio is called the *parameter* of the point. Note that the parameter of Q is 0, while the parameter of P is said to be ∞. Thus the real numbers can be put into one-to-one correspondence with all points on a line except for one, namely, the first base point.

Using homogeneous parameters and Theorem 4.19, we can now show that our analytic model satisfies Axiom 5. (If two triangles are perspective from a point, then they are perspective from a line.)

Verification of Axiom 5

Let the two triangles have vertices $A(a_1, a_2, a_3)$, $B(b_1, b_2, b_3)$, $C(c_1, c_2, c_3)$ and $A'(a'_1, a'_2, a'_3)$, $B'(b'_1, b'_2, b'_3)$, $C'(c'_1, c'_2, c'_3)$. Assume these triangles are perspective from $P(p_1, p_2, p_3)$. Let $Q = AB \cdot A'B'$, $R = BC \cdot B'C'$, and $S = AC \cdot A'C'$. We need to show that Q, R, and S are collinear (Fig. 4.34).

To do this we make use of homogeneous parameters. Since P is on line AA', BB', and CC', it has homogeneous parameters $(\alpha_1, \alpha_2), (\beta_1, \beta_2)$, and (γ_1, γ_2) with respect to base points A and A', B and B', C and C', respectively. Thus the homogeneous coordinates of P are given by $p_i = \alpha_1 a_i + \alpha_2 a'_i = \beta_1 b_i + \beta_2 b'_i = \gamma_1 c_i + \gamma_2 c'_i$, $i = 1, 2, 3$. The first two of these yield $\alpha_1 a_i - \beta_1 b_i = \beta_2 b'_i - \alpha_2 a'_i$ so $(\alpha_1 a_1 - \beta_1 b_1, \alpha_1 a_2 - \beta_1 b_2, \alpha_1 a_3 - \beta_1 b_3) = (\beta_2 b'_1 - \alpha_2 a'_1, \beta_2 b'_2 - \alpha_2 a'_2, \beta_2 b'_3 - \alpha_2 a'_3)$,

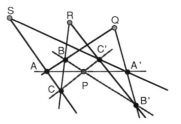

Figure 4.34

but the first of these ordered triples gives homogeneous coordinates for a point on line AB whereas the second gives homogeneous coordinates for a point on line $A'B'$. Since the two triples are equal, both must be coordinates for the point Q. We will use the first set. Likewise we can show that $R(\beta_1 b_1 - \gamma_1 c_1, \beta_1 b_2 - \gamma_1 c_2, \beta_1 b_3 - \gamma_1 c_3)$ and finally that $S(\alpha_1 a_1 - \gamma_1 c_1, \alpha_1 a_2 - \gamma_1 c_2, \alpha_1 a_3 - \gamma_1 c_3)$. Using these homogeneous coordinates, we can show that $|QRS| = 0$ so that the three points are indeed collinear.

EXERCISES

1. Show that the interpretation of incidence used in our analytic model is independent of the particular homogeneous coordinates used for the point X and the line u.

2. Verify that the set of interpretations of the undefined terms given in this section satisfies Axioms 1–3.

3. (a) Describe the homogeneous coordinates of the ideal points. (b) Find homogeneous coordinates of the ideal line. (c) Show analytically that the ideal points lie on the ideal line.

4. Let $l[a, b, c]$ be a line. Find homogeneous coordinates of the ideal point(s) on l. How many ideal points are there on l?

5. (a) Find the equation of the line joining the points $(0, 2, 1)$ and $(1, 1, 0)$. (b) Find a set of coordinates for this line. (c) Find the point of intersection of the lines $2x_2 + x_3 = 0$ and $x_1 + x_2 = 0$. (d) Find the line joining the points $2u_2 + u_3 = 0$ and $u_1 + u_2 = 0$.

6. Let $l[a, b, c]$ and $m[a, b, d]$ be two distinct lines in the projective plane (i.e., $c \neq d$). (a) Find the point of intersection of l and m. (b) Do l and m intersect in the Euclidean plane? Why?

7. Let X, Y, Z be the points with homogeneous coordinates $(1, 0, 0)$, $(0, 1, 0)$, and $(0, 0, 1)$, respectively. (a) Show that X, Y, Z are noncollinear. (b) Show that if $P(p_1, p_2, p_3)$ is any point distinct from Z then the point $P' = ZP \cdot XY$ has homogeneous coordinates $(p_1, p_2, 0)$.

8. Show that the points $P(2, 3, -2)$, $Q(1, 2, -4)$, and $R(0, 1, -6)$ are collinear and find homogeneous parameters of R with respect to P and Q. What is the corresponding parameter of R?

9. Find an example showing that the homogeneous parameters of a point with respect to a given pair of base points depend on the homogeneous coordinates used for the base points.

10. Use ordered triples consisting of 0's and 1's and arithmetic modulo 2 to coordinatize the finite projective plane with three points on a line (see Section 1.3).

4.8. The Analytic Form of Projectivities

Using our analytic model of the projective plane, it is now possible to find 2×2 matrix representations of the one-to-one correspondences between the elements of two pencils known as projectivities.

Theorem 4.20. *A projectivity between the elements of two pencils can be represented by a real matrix equation of the form*

$$s\begin{bmatrix} \lambda_1' \\ \lambda_2' \end{bmatrix} = \begin{bmatrix} a_{11} & a_{12} \\ a_{21} & a_{22} \end{bmatrix}\begin{bmatrix} \lambda_1 \\ \lambda_2 \end{bmatrix}$$

where $a_{11}a_{22} - a_{21}a_{12} = |A| \neq 0$, $s \neq 0$, *and where* (λ_1, λ_2) *and* (λ_1', λ_2') *are homogeneous parameters of the original and image elements with respect to predetermined base elements.*

Proof. We first show that a perspectivity from a pencil of points to a pencil of lines has this algebraic form.

Let P and Q be base points of the pencil of points and let the lines m and n be base lines of the pencil of lines. Let $X(\lambda_1, \lambda_2)$ be any other point on line PQ; assume it corresponds to line $x'(\lambda_1', \lambda_2')$. By the definition of the perspectivity, $x' \cdot X = 0$. Writing this out in terms of components, gives the following equation:

$$[\lambda_1' m_1 + \lambda_2' n_1, \lambda_1' m_2 + \lambda_2' n_2, \lambda_1' m_3 + \lambda_2' n_3]$$
$$\cdot (\lambda_1 p_1 + \lambda_2 q_1, \lambda_1 p_2 + \lambda_2 q_2, \lambda_1 p_3 + \lambda_2 q_3) = 0$$

or

$$\lambda_1' \lambda_1 (p_1 m_1 + p_2 m_2 + p_3 m_3) + \lambda_1' \lambda_2 (q_1 m_1 + q_2 m_2 + q_3 m_3)$$
$$+ \lambda_2' \lambda_1 (p_1 n_1 + p_2 n_2 + p_3 n_3) + \lambda_2' \lambda_2 (q_1 n_1 + q_2 n_2 + q_3 n_3) = 0 \quad (4.1)$$

To simplify Equation (4.1), we use a substitution for each of the sums in parentheses:

$$a_{21} = \sum p_i m_i, \quad a_{22} = \sum q_i m_i, \quad a_{11} = -\sum p_i n_i, \quad a_{12} = -\sum q_i n_i$$

so

$$a_{21}\lambda_1' \lambda_1 + a_{22}\lambda_1' \lambda_2 - a_{11}\lambda_2' \lambda_1 - a_{12}\lambda_2' \lambda_2 = 0 \quad (4.2)$$

or

$$\lambda_1'(a_{21}\lambda_1 + a_{22}\lambda_2) = \lambda_2'(a_{11}\lambda_1 + a_{12}\lambda_2)$$

This gives

$$\frac{\lambda_1'}{\lambda_2'} = \frac{a_{11}\lambda_1 + a_{12}\lambda_2}{a_{21}\lambda_1 + a_{22}\lambda_2}$$

which can be written in matrix notation as

$$s\begin{bmatrix} \lambda_1' \\ \lambda_2' \end{bmatrix} = \begin{bmatrix} a_{11} & a_{12} \\ a_{21} & a_{22} \end{bmatrix}\begin{bmatrix} \lambda_1 \\ \lambda_2 \end{bmatrix}$$

where $s \neq 0$. Note that $a_{11}a_{22} - a_{21}a_{12} = |A| \neq 0$ since the projectivity is a one-to-one mapping.

If we solve Equation (4.2) for λ_1/λ_2, we obtain a similar representation for a perspectivity from a pencil of lines to a pencil of points. Since any projectivity is a finite product of perspectivities and any perspectivity either maps from a pencil of points to a pencil of lines, from a pencil of lines to a pencil of points, or is a product of these two, it is sufficient to note that the product of two matrices of this form will also be a matrix of this form.   □

Using the matrix representation given in Theorem 4.20, it is now relatively easy to verify Axiom 6 (see Exercise 3). However, the verification of this axiom, as well as any other use of the matrix equation for a projectivity, requires careful attention to s, the scalar involved. Since the homogeneous parameters of elements of pencils are not unique, it is essential to allow s to take on different values even within the context of a given projectivity. The following example illustrates the way in which this indeterminate nature of the scalar must be handled in finding the matrix of a particular projectivity.

EXAMPLE 4.2. Find a matrix of the projectivity that maps points on p with homogeneous parameters $(1, 3)$, $(1, 2)$, and $(2, 3)$ to points on p' with homogeneous parameters $(1, -4)$, $(0, 1)$, and $(-1, 1)$, respectively.

By Theorem 4.20, the projectivity can be represented by a 2×2 matrix A where

$$s \begin{bmatrix} \lambda_1' \\ \lambda_2' \end{bmatrix} = \begin{bmatrix} a & b \\ c & d \end{bmatrix} \begin{bmatrix} \lambda_1 \\ \lambda_2 \end{bmatrix}$$

The algebra involved in finding such a matrix can be simplified somewhat by first considering any cases where the homogeneous parameters of either a point or its image include the value 0. In this case we will first impose the condition that the ordered pair $(1, 2)$ map to $(0, 1)$. It is also helpful to find and use all convenient substitutions as soon as possible. We will make use of both techniques in the following calculations.

In order to map $(1, 2)$ to $(0, 1)$ we must have

$$s_1 \begin{bmatrix} 0 \\ 1 \end{bmatrix} = \begin{bmatrix} a & b \\ c & d \end{bmatrix} \begin{bmatrix} 1 \\ 2 \end{bmatrix} \quad \text{or} \quad \begin{array}{ll} 0 = a + 2b & (4.3) \\ s_1 = c + 2d & (4.4) \end{array}$$

Equation (4.3) yields $a = -2b$, so we can use this substitution when requiring that the matrix map $(1, 3)$ to $(1, -4)$.

$$s_2 \begin{bmatrix} 1 \\ -4 \end{bmatrix} = \begin{bmatrix} -2b & b \\ c & d \end{bmatrix} \begin{bmatrix} 1 \\ 3 \end{bmatrix} \quad \text{or} \quad \begin{array}{ll} s_2 \doteq b & (4.5) \\ -4s_2 = c + 3d & (4.6) \end{array}$$

Equation (4.5) allows us to replace s_2 with b in equation (4.6), giving $c = -3d - 4b$. Using this subsitution, the third point and its image give the following equations:

$$s_3 \begin{bmatrix} -1 \\ 1 \end{bmatrix} = \begin{bmatrix} -2b & b \\ -3d-4b & d \end{bmatrix} \begin{bmatrix} 2 \\ 3 \end{bmatrix} \quad \text{or} \quad \begin{array}{l} -s_3 = -b \qquad (4.7) \\ s_3 = -3d-8b \qquad (4.8) \end{array}$$

Since Equations (4.7) and (4.8) involve the three unknowns b, d, and s_3; we can choose a value for one of the unknowns. Let $s_3 = 1$. Then Equation (4.7) gives $b = 1$ and Equation (4.8) gives $d = -3$. Using these values in Equations (4.5) and (4.6) yields $c = 5$, and finally Equation (4.3) gives $a = -2$. So the matrix A is the matrix

$$\begin{bmatrix} -2 & 1 \\ 5 & -3 \end{bmatrix}$$

Note that the scalar s did assume different values, namely, $s_1 = -1$, while $s_2 = s_3 = 1$.

The following proof of Theorem 4.21 (the converse of Theorem 4.20) also illustrates the way in which the matrix of a projectivity is determined.

Theorem 4.21. *Any mapping given by an equation of the following form is a projectivity:*

$$s \begin{bmatrix} \lambda'_1 \\ \lambda'_2 \end{bmatrix} = \begin{bmatrix} a & b \\ c & d \end{bmatrix} \begin{bmatrix} \lambda_1 \\ \lambda_2 \end{bmatrix} \qquad ad - bc \neq 0, \quad s \neq 0 \qquad (4.9)$$

Proof. The proof assumes that both pencils are pencils of points; however, identical arguments can be made for the other cases. Let $P(1,0)$ and $Q(0,1)$ be the base points for the first pencil of points. Let R be the point with parameters $(1,1)$ with respect to P and Q. Then under the mapping given by this matrix equation, $P'(a,c)$, $Q'(b,d)$, and $R'(a+b,c+d)$ are the corresponding elements of the second pencil with respect to a predetermined basis. By the fundamental theorem, there is a unique projectivity T such that $T : PQR \to P'Q'R'$, but by Theorem 4.20 the projectivity T has a matrix equation

$$s \begin{bmatrix} \lambda'_1 \\ \lambda'_2 \end{bmatrix} = \begin{bmatrix} a_{11} & a_{12} \\ a_{21} & a_{22} \end{bmatrix} \begin{bmatrix} \lambda_1 \\ \lambda_2 \end{bmatrix}$$

It is then sufficient to show that this matrix is a scalar multiple of the matrix in Equation (4.9). To evaluate a, b, c, d, we will determine the algebraic conditions necessary for mapping P to P', Q to Q', and R to R'. Since the scalar s may differ from point to point, we need to allow s to assume different values in each of these cases leading to the following three equations:

$$s_1 \begin{bmatrix} a \\ c \end{bmatrix} = \begin{bmatrix} a_{11} & a_{12} \\ a_{21} & a_{22} \end{bmatrix} \begin{bmatrix} 1 \\ 0 \end{bmatrix} \qquad s_2 \begin{bmatrix} b \\ d \end{bmatrix} = \begin{bmatrix} a_{11} & a_{12} \\ a_{21} & a_{22} \end{bmatrix} \begin{bmatrix} 0 \\ 1 \end{bmatrix}$$

$$s_3 \begin{bmatrix} a+b \\ c+d \end{bmatrix} = \begin{bmatrix} a_{11} & a_{12} \\ a_{21} & a_{22} \end{bmatrix} \begin{bmatrix} 1 \\ 1 \end{bmatrix}$$

These matrix equations yield the following:

$$s_1 a = a_{11} \qquad s_2 b = a_{12} \qquad s_3(a + b) = a_{11} + a_{12}$$
$$s_1 c = a_{21} \qquad s_2 d = a_{22} \qquad s_3(c + d) = a_{21} + a_{22}$$

Since there are six equations in seven unknowns, one unknown can be chosen. Chose $s_3 = 1$. Then

$$a + b = a_{11} + a_{12} = s_1 a + s_2 b$$
$$c + d = a_{21} + a_{22} = s_1 c + s_2 d$$

so

$$a(1 - s_1) + b(1 - s_2) = 0$$
$$c(1 - s_1) + d(1 - s_2) = 0$$

and since $ad - bc \neq 0$, the solution $s_1 = 1, s_2 = 1$ is unique. Thus $a = a_{11}$ and so on, and the matrix equation

$$s \begin{bmatrix} \lambda_1' \\ \lambda_2' \end{bmatrix} = \begin{bmatrix} a_{11} & a_{12} \\ a_{21} & a_{22} \end{bmatrix} \begin{bmatrix} \lambda_1 \\ \lambda_2 \end{bmatrix}$$

is the representation of a projectivity.   ☐

Together Theorems 4.20 and 4.21 tell us that there is a one-to-one correspondence between the set of projectivities between two pencils relative to predetermined base elements and the set of equivalence clases of 2×2 matrices with nonzero determinants where $A \sim B$ iff $A = sB$ for some nonzero constant s.

According to Axiom 6, projectivities on a pencil other than the identity have two or fewer invariant elements. The next theorem characterizes the matrix representations of projectivities with two, one, and zero invariant elements, respectively. The proof of this theorem makes use of an important result about eigenvectors from linear algebra.

Theorem 4.22. *A projectivity on a pencil, other than the identity, with matrix*

$$\begin{bmatrix} a_{11} & a_{12} \\ a_{21} & a_{22} \end{bmatrix}$$

has two distinct invariant elements, one invariant element, or no invariant elements according as $(a_{22} - a_{11})^2 + 4a_{12}a_{21} > 0$, $= 0$, or < 0.

Proof. Note (λ_1, λ_2) is an invariant element iff

$$s \begin{bmatrix} \lambda_1 \\ \lambda_2 \end{bmatrix} = \begin{bmatrix} a_{11} & a_{12} \\ a_{21} & a_{22} \end{bmatrix} \begin{bmatrix} \lambda_1 \\ \lambda_2 \end{bmatrix}$$

that is, iff (λ_1, λ_2) is a characteristic vector or eigenvector of the matrix. But eigenvectors exist iff there is a nonzero solution of the characteristic equation $|A - sI| = 0$. Evaluating $|A - sI|$ gives $(a_{11} - s)(a_{22} - s) - a_{12}a_{21} = 0$. Expanding and solving for s yields the following:

$$s = \frac{(a_{22} + a_{11}) \pm \sqrt{(a_{22} + a_{11})^2 - 4(a_{11}a_{22} - a_{12}a_{21})}}{2}.$$

If the expression under the radical is positive, there are two distinct solutions for s and therefore two linearly independent eigenvectors and hence two distinct invariant points of the projectivity. If this expression is zero, there is exactly one solution for s and therefore exactly one invariant point of the projectivity (see Exercise 5). Finally, if the expression is negative, there are no real valued solutions for s and so no invariant points of the projectivity. Since this expression is algebraically eqivalent to the expression in the statement of the theorem, the result follows. ☐

Definition 4.17. A projectivity on a pencil is called *hyperbolic, parabolic,* or *elliptic* if the number of invariant elements is 2, 1, or 0, respectively.

These definitions are suggestive of a connection between projectivities on pencils and similarly named conics, which will be formalized in Section 4.12.

EXERCISES

1. If under a projectivity between pencils the base elements P and Q of the first pencil correspond to the base elements P' and Q' of the second pencil, respectively, show that the matrix of the projectivity is a diagonal matrix.

2. Find the matrix of the projectivity that maps points on p with homogeneous parameters $(0, 1)$, $(1, 0)$, and $(1, 1)$ to points on p' with homogeneous parameters $(1, 2)$, $(2, 3)$, and $(-1, 0)$, respectively.

3. Use Theorem 4.20 to verify that our analytic model satisfies Axiom 6. [*Hint:* You may want to choose two of the invariant elements as base elements.]

4. Find the homogeneous parameters of the invariant elements under the projectivity on a pencil that has the following matrix representation:

$$\begin{bmatrix} 6 & -4 \\ 1 & 1 \end{bmatrix}$$

5. A result from linear algebra says that there is at least one eigenvector corresponding to each solution s of the equation $|A - sI| = 0$. Show that there cannot be two linearly independent eigenvectors corresponding to the same solution s when A is a 2×2 matrix.

The following exercises refer to a special type of projectivity known as an involution. An *involution* is a transformation $T \neq I$ such that $T^2 = I$.

6. Prove that a projectivity on a pencil that interchanges one pair of distinct

elements is an involution. [*Hint:* Use the two points that are interchanged as base points and find the matrix representation.]

7. Show that in general the matrix of an involution is of the form

$$\begin{bmatrix} a & b \\ c & -a \end{bmatrix} \quad \text{where } a^2 + bc \neq 0.$$

8. Show that an involution with a matrix of the form given in Exercise 7 is elliptic iff $a^2 + bc < 0$.

9. By using two points that are interchanged as base points, show that the matrix of an elliptic involution is of the form given in Exercise 7 with $a = 0$, and $bc < 0$.

4.9. Cross Ratios

Within the context of the analytic model of the projective plane, it is natural to ask if the Euclidean concept of distance is relevant. However, as previously indicated, projective geometry studies the invariants under transformations that can be considered as generalized affinities and these in turn are generalizations of similarities. In Chapter 3, we discovered that similarities do not preserve distances but only ratios of distances, and affinities only preserve segment division ratios. This suggests that the concept of distance may not be relevant in projective geometry, so it is surprising that we are able to show that projective transformations do preserve a numerical value called the cross ratio, which can actually be interpreted as a ratio of ratios of distances.

Definition 4.18. If A, B, C, D are four distinct elements of a pencil with homogeneous parameters (α_1, α_2), (β_1, β_2), (γ_1, γ_2), and (δ_1, δ_2) with respect to given base points, then the *cross ratio* of the four elements, in the given order, is the number given by the following equation involving determinants:

$$R(A, B, C, D) = \frac{\begin{vmatrix} \gamma_1 & \alpha_1 \\ \gamma_2 & \alpha_2 \end{vmatrix}}{\begin{vmatrix} \gamma_1 & \beta_1 \\ \gamma_2 & \beta_2 \end{vmatrix}} \div \frac{\begin{vmatrix} \delta_1 & \alpha_1 \\ \delta_2 & \alpha_2 \end{vmatrix}}{\begin{vmatrix} \delta_1 & \beta_1 \\ \delta_2 & \beta_2 \end{vmatrix}}.$$

If none of the four elements A, B, C, D is the first base element, each also has homogeneous parameters $(\alpha, 1)$, $(\beta, 1)$, $(\gamma, 1)$, $(\delta, 1)$, respectively, where α, β, γ, δ are the corresponding (nonhomogeneous) parameters. Note in this case

$$R(A, B, C, D) = \frac{\gamma - \alpha}{\gamma - \beta} \div \frac{\delta - \alpha}{\delta - \beta}.$$

It is this restatement of the definition that makes the interpretation of the definition as a ratio of ratios of distances more apparent (see Exercise 3).

Even though the notation used in the previous definition is suggestive of a pencil of points, the definition applies to both pencils of points and pencils of lines. We will make use of similar notation in the following theorems, which indicate how changes in the order of the elements affect the cross ratio. The proofs of these theorems follow from the definition by algebraic computation.

Theorem 4.23. *If A, B, C, D are four distinct elements of a pencil, then the cross ratio $R(A, B, C, D)$ remains unchanged when any two pairs of the elements are interchanged; that is, $R(A, B, C, D) = R(B, A, D, C) = R(C, D, A, B) = R(D, C, B, A)$.*

Theorem 4.24. *If the cross ratio of four distinct elements of a pencil named in a given order is r, interchanging either the first or second pair of elements changes the cross ratio to its reciprocal $1/r$, interchanging either the inner pair or the outer pair changes the cross ratio r to $1 - r$.*

Corollary. *The 24 possible permutations of four distinct elements of a pencil can be categorized into six sets of four, corresponding to cross ratios of r, $1/r$, $1 - r$, $(r - 1)/r$, $r/(r - 1)$, and $1/(1 - r)$.*

Theorem 4.25. *The cross ratio of four distinct elements of a pencil cannot be 0, 1, or ∞.*

Since projective transformations will induce projectivities mapping one pencil to another, demonstrating the invariance of the cross ratio under projectivities will verify its invariance under projective transformations.

Theorem 4.26. *The cross ratio of four distinct elements of a pencil is invariant under a projectivity (so, e.g., if $ABCD \wedge A'B'C'D'$, then $R(A, B, C, D) = R(A', B', C', D')$).*

Proof. Assume that distinct elements A, B, C, D of one pencil map to corresponding elements A', B', C', D' of a second pencil under a projectivity with matrix $A = [a_{ij}]$. Then

$$\begin{vmatrix} \gamma_1' & \alpha_1' \\ \gamma_2' & \alpha_2' \end{vmatrix} = \begin{vmatrix} a_{11} & a_{12} \\ a_{21} & a_{22} \end{vmatrix} \begin{vmatrix} \gamma_1 & \alpha_1 \\ \gamma_2 & \alpha_2 \end{vmatrix}$$

where A has homogeneous parameters (α_1, α_2), A' has homogeneous parameters (α_1', α_2'), and so on, with respect to predetermined base elements. So

$$R(A', B', C', D') = \frac{\begin{vmatrix} \gamma_1' & \alpha_1' \\ \gamma_2' & \alpha_2' \end{vmatrix}}{\begin{vmatrix} \gamma_1' & \beta_1' \\ \gamma_2' & \beta_2' \end{vmatrix}} \div \frac{\begin{vmatrix} \delta_1' & \alpha_1' \\ \delta_2' & \alpha_2' \end{vmatrix}}{\begin{vmatrix} \delta_1' & \beta_1' \\ \delta_2' & \beta_2' \end{vmatrix}}$$

$$= \frac{\begin{vmatrix} a_{11} & a_{12} \\ a_{21} & a_{22} \end{vmatrix} \begin{vmatrix} \gamma_1 & \alpha_1 \\ \gamma_2 & \alpha_2 \end{vmatrix}}{\begin{vmatrix} a_{11} & a_{12} \\ a_{21} & a_{22} \end{vmatrix} \begin{vmatrix} \gamma_1 & \beta_1 \\ \gamma_2 & \beta_2 \end{vmatrix}} \div \frac{\begin{vmatrix} a_{11} & a_{12} \\ a_{21} & a_{22} \end{vmatrix} \begin{vmatrix} \delta_1 & \alpha_1 \\ \delta_2 & \alpha_2 \end{vmatrix}}{\begin{vmatrix} a_{11} & a_{12} \\ a_{21} & a_{22} \end{vmatrix} \begin{vmatrix} \delta_1 & \beta_1 \\ \delta_2 & \beta_2 \end{vmatrix}}$$

$$= \frac{\begin{vmatrix} \gamma_1 & \alpha_1 \\ \gamma_2 & \alpha_2 \\ \gamma_1 & \beta_1 \\ \gamma_2 & \beta_2 \end{vmatrix}}{\begin{vmatrix} \delta_1 & \alpha_1 \\ \delta_2 & \alpha_2 \\ \delta_1 & \beta_1 \\ \delta_2 & \beta_2 \end{vmatrix}}$$

$$= R(A, B, C, D). \qquad \qquad \square$$

This theorem leads to a useful corollary, which enables the computation of the cross ratio of four elements directly from the homogeneous coordinates of the elements rather than from homogeneous parameters, which in turn must be first computed relative to given base points.

Corollary. *If A, B, C, D, with homogeneous coordinates (a_1, a_2, a_3) and so on, are four distinct elements of a pencil not containing $Z(0, 0, 1)$ then*

$$\frac{\begin{vmatrix} c_1 & a_1 \\ c_2 & a_2 \\ c_1 & b_1 \\ c_2 & b_2 \end{vmatrix}}{\begin{vmatrix} d_1 & a_1 \\ d_2 & a_2 \\ d_1 & b_1 \\ d_2 & b_2 \end{vmatrix}} = R(A, B, C, D)$$

Thus if $Z(0, 0, 1)$ is not an element of the pencil, the first two homogeneous coordinates of the elements can be used in the role of homogeneous parameters in the cross ratio. But if the pencil does contain $Z(0, 0, 1)$ (so this corollary fails to hold), then it cannot also contain both $X(1, 0, 0)$ and $Y(0, 1, 0)$, and comparable corollaries can be proved for pencils not containing X and for pencils not containing Y (see Exercise 8). The use of one of these comparable statements is demonstrated in Example 4.3.

EXAMPLE 4.3. Find $R(A, B, C, D)$ where $A(1, 2, 1)$, $B(3, 6, 1)$, $C(2, 4, 1)$, and $D(1, 2, 0)$ are points on $l[2, -1, 0]$. (Note that if these points are identified as points in the Euclidean plane, C would be called the midpoint of segment AB.)

Since $Z(0, 0, 1)$ is clearly a point on l, we cannot use the corollary to Theorem 4.26 directly. However, since $X(1, 0, 0)$ is not incident with l, we can use a comparable result; that is, we can use the last two homogeneous coordinates of each point in the role of homogeneous parameters in order to compute the cross ratio. This gives

$$R(A, B, C, D) = \frac{\begin{vmatrix} 4 & 2 \\ 1 & 1 \\ 4 & 6 \\ 1 & 1 \end{vmatrix}}{\begin{vmatrix} 2 & 2 \\ 0 & 1 \\ 2 & 6 \\ 0 & 1 \end{vmatrix}} = \frac{2}{-2} \div \frac{2}{2} = -1$$

Recall that the fundamental theorem of projective geometry indicates that in general there exists a projectivity mapping any three elements of one pencil to any three corresponding elements of a second pencil. However, as will be shown in Theorem 4.28, if any four elements of the first pencil are named and any four corresponding elements of the second pencil *with the same cross ratio* are given, there is a projectivity mapping the first set of four elements to the second set of four elements. The proof of this result requires one additional property of cross ratios that can also be verified by algebraic computation.

Theorem 4.27. *If three distinct elements of a pencil, A, B, C, and a real number r $(r \neq 0, 1)$ are given, then there exists a unique point D such that $R(A, B, C, D) = r$.*

Theorem 4.28. *If A, B, C, D are four distinct elements of one pencil and A', B', C', D' are four distinct elements of a second pencil, with $R(A', B', C', D') = R(A, B, C, D)$, then there exists a projectivity mapping A, B, C, D to A', B', C', D', respectively.*

Proof. By the fundamental theorem, there exists a projectivity such that $ABC \wedge A'B'C'$. Let D^* be the unique image of D under this projectivity. By Theorem 4.26 $R(A, B, C, D) = R(A', B', C', D^*)$; but $R(A, B, C, D) = R(A', B', C', D')$. So $D^* = D'$. $\qquad\qquad\Box$

The proof of Theorem 4.28, together with the previous theorems indicating the changes in the cross ratio resulting from various possible changes in the ordering of the four elements, is reminiscent of similar theorems about harmonic sets and suggests a possible relation between the two concepts. This relation is formalized in the final theorem of this section.

Theorem 4.29. *If A, B, C, D are four distinct elements of a pencil, then $R(A, B, C, D) = -1$ iff $H(AB, CD)$.*

Proof. (a) Since $H(AB, CD)$, it follows that $H(AB, DC)$ and, by Theorem 4.10, there is a projectivity such that $ABCD \wedge ABDC$. Thus by Theorem 4.26 $R(A, B, C, D) = R(A, B, D, C)$; but by Theorem 4.24, if $R(A, B, C, D) = r$, then $R(A, B, D, C) = 1/r$. Thus $r = 1/r$ or $r^2 = 1$. Since $r \neq 1$, this implies that $r = -1$.

(b) Assume $R(A, B, C, D) = -1$. Let D' be a fourth element of a pencil such that $H(AB, CD')$. Thus by the previous part of the proof, $R(A, B, C, D') = -1$, and it follows by Theorem 4.27 that $D = D'$. $\qquad\qquad\Box$

EXERCISES

1. Given collinear points with their homogeneous parameters, $A(1, 1)$, $B(3, 2)$, $C(1, 0)$, $D(-1, 2)$, find $R(A, B, C, D)$ and $R(C, A, B, D)$.

2. Find the coordinates of a point D that is collinear with $A(3, 1, 2)$, $B(1, 0, -1)$, $C(1, 1, 4)$ with $R(A, B, C, D) = -\frac{2}{3}$.

3. In the Euclidean plane, let A, B, C, D be distinct points on a number line with coordinates α, β, γ, δ, respectively. Show that the (nonhomogeneous) parameter form of the cross ratio $R(A, B, C, D)$ is the ratio of two segment division ratios of these four points (see Section 3.9).

4. Show that if C has homogeneous parameters $(1, 1)$ and D has homogeneous parameters $(r, 1)$ with respect to A and B, then $R(A, B, C, D) = r$.

5. Prove Theorem 4.24.

6. Prove Theorem 4.25.

7. Prove the corollary to Theorem 4.26. [*Hint*: See Exercise 7 in Section 4.7.]

8. What is the comparable statement to the corollary to Theorem 4.26 for pencils not containing $X(1, 0, 0)$? For pencils not containing $Y(0, 1, 0)$?

9. Prove Theorem 4.27.

10. Prove: If A, B, C, D, E are five distinct collinear points, then $R(A, B, C, D) \cdot R(A, B, D, E) = R(A, B, C, E)$.

4.10. Collineations

There are two distinct types of transformations of the projective plane. The transformations considered in this section map collinear points to collinear points (and thus lines to lines). These transformations, called *collineations*, form a group; it is the invariants of this group that are studied in projective geometry. In the next section, we consider transformations that map collinear points to concurrent lines (and thus lines to points). These transformations, called *correlations*, allow mappings between dual figures and provide an analytic equation for conics.

If we let V be the set of points of the analytic model of the projective plane together with $\{(0, 0, 0)\}$ (i.e., V is the set of *all* equivalence classes of R^3), we can show that V is a vector space under the usual addition and scalar multiplication in R^3 (see Exercise 1). Collineations are defined as linear transformations of this vector space.

Definition 4.19. A one-to-one linear transformation of V onto itself is a *collineation*.

With this definition, a slight modification of Theorem 3.3 leads to the following result giving the analytic form for collineations (see Exercise 2).

Theorem 4.30. *A collineation can be represented by a 3×3 real valued matrix A where $|A| \neq 0$. The matrix equation for the collineation is $sX' = AX$ where $X \in R^3$ and $s \neq 0$.*

There are two important observations about this theorem we should make. First, equations of collineations, like equations of projectivities, contain nonzero scalars, and it is essential to allow this scalar to take on different values even within the context of the same collineation. Second, the matrix of a collineation is not unique (since if A is the matrix of a given collineation, kA will also be a matrix of the collineation for any nonzero scalar k), but there is a unique equivalence class of matrices corresponding to each collineation (see Exercise 4).

To show that the term "collineation" is appropriate, we need to verify that these mappings do indeed preserve collinearity as previously claimed. We can then conclude that collineations induce mappings from lines to lines, so it is appropriate to look for an equation that gives the image of a line directly.

Theorem 4.31. *A collineation maps collinear points to collinear points. The image of a line $u[u_1, u_2, u_3]$ under a collineation with matrix A is given by the equation $ku' = uA^{-1}$, $k \neq 0$.*

Proof. Assume that P is a point on line QR. Then it suffices to show that P', the image of P under the collineation, is collinear with the images of Q and R, namely, Q' and R'. Since P is on QR, Theorem 4.19 implies that there are two real numbers λ_1 and λ_2 such that $P = \lambda_1 Q + \lambda_2 R$. Then $sP' = AP = A(\lambda_1 Q + \lambda_2 R)$ for some nonzero scalar s, or $P' = (\lambda_1/s)AQ + (\lambda_2/s)AR = \lambda_1 Q' + \lambda_2 R'$ so, again by Theorem 4.19, P' is on line $Q'R'$.

To find the equation of the image line, assume the collineation with matrix A maps the line with coordinates u and equation $uX = 0$ to the line with coordinates u' and equation $u'X' = 0$ where $sX' = AX$ for some nonzero scalar s. Replacing X' in the equation of the image line with $(1/s)AX$ yields $u'X' = (1/s)u'AX = 0$. Thus the point X' is on the line $u'X' = 0$ iff X is on the line $(u'A)X = 0$, but X' is on $u'X' = 0$ iff X is on $uX = 0$. Since collineations are one-to-one mappings, $u'AX = 0$ and $uX = 0$ must be the same line. Thus $u = ku'A$ or $ku' = uA^{-1}$. $\qquad\qquad\square$

This means that the *same* collineation that maps points according to the equation $sX' = AX$ maps lines according to the equation $ku' = uA^{-1}$. Thus there are two equations that describe the mapping of any particular collineation: a *point equation* that gives the images of points and a *line equation* that gives the images of lines. The matrix A used in the point equation is called the *matrix of the collineation*.

Since a collineation maps collinear points to collinear points, duality suggests that it will also map concurrent lines to concurrent lines. The proof of this corollary begins with the line equation of the collineation and is an exact parallel of the first part of the proof of Theorem 4.31.

Corollary. *Under a collineation, concurrent lines are mapped to concurrent lines.*

The set of collineations under the operation of composition form a group as can be verified by using the definition of a group (see Exercise 5).

Theorem 4.32. *The set of collineations forms a group under composition.*

As noted several times previously, Klein defined projective geometry as the study of properties of V that are invariant under the group of collineations. The following theorem shows that the properties of cross ratio and harmonic relation, which we have previously shown invariant under projectivities, are also invariant under collineations.

Theorem 4.33. *A collineation of the projective plane induces a projectivity between the elements of corresponding pencils.*

Proof. Let P, Q, R be three collinear points, so $R = \lambda_1 P + \lambda_2 Q$. Let P', Q', and R' be their images under a collineation with matrix A. Then P', Q', and R' are also collinear so $R' = \mu_1 P' + \mu_2 Q'$. Applying the collineation to P, Q, and R yields $s_1 P' = AP$, $s_2 Q' = AQ$, $s_3 R' = AR$, where each $s_i \neq 0$. Since $R = \lambda_1 P + \lambda_2 Q$, this last equation gives $s_3 R' = A(\lambda_1 P + \lambda_2 Q) = \lambda_1 AP + \lambda_2 AQ = s_1 \lambda_1 P' + s_2 \lambda_2 Q'$. So when P and Q are base elements of the first pencil and P' and Q' are base elements of the second pencil, the element R of the first pencil has homogeneous parameters (λ_1, λ_2), whereas its image R' has homogeneous parameters (μ_1, μ_2) where

$$s_3 \begin{bmatrix} \mu_1 \\ \mu_2 \end{bmatrix} = \begin{bmatrix} s_1 & 0 \\ 0 & s_2 \end{bmatrix} \begin{bmatrix} \lambda_1 \\ \lambda_2 \end{bmatrix} \quad \text{and} \quad \begin{vmatrix} s_1 & 0 \\ 0 & s_2 \end{vmatrix} \neq 0.$$

Therefore, by Theorem 4.21 the induced mapping between pencils of points is a projectivity. The proof for pencils of lines follows by duality. ☐

Corollary. Cross ratios and harmonic sets are invariant under collineations.

Having established the connection between collineations of the projective plane and projectivities of pencils, we will now study the general properties of collineations. Whereas projectivities are uniquely determined by three pairs of corresponding elements, the next theorem shows that collineations are uniquely determined by four pairs of corresponding elements. The proof of this theorem illustrates a useful technique for finding the matrix of a collineation.

Theorem 4.34. *There exists a unique collineation that maps any four points, no three collinear, to any four points, no three collinear.*

Proof. The verification of this theorem consists of algebraically finding a matrix A of the collineation that maps any four points P, Q, R, S (no three collinear) to any four points P', Q', R', S' (no three collinear) and noting that this matrix is uniquely determined modulo the equivalence relation. This procedure can be simplified somewhat by first finding a matrix B such that $s_1 P' = BX$, $s_2 Q' = BY$, $s_3 R' = BZ$, and $s_4 S' = BU$ where $X(1, 0, 0)$, $Y(0, 1, 0)$, $Z(0, 0, 1)$, and $U(1, 1, 1)$, and then finding a matrix C such that $s_5 P = CX$, $s_6 Q = CY$, $s_7 R = CZ$, $s_8 S = CU$. The matrix A is then given by $A = BC^{-1}$.

Corollary. *A collineation of the plane with four invariant points, no three collinear, is the identity transformation.*

Clearly a collineation is also uniquely determined by four lines (no three concurrent) and four image lines (no three concurrent) and the matrix A^{-1} used in the line equation of the collineation can be found by a procedure similar to that outlined in the proof of Theorem 4.34. This "simplified" procedure for finding the matrix of a collineation that maps a given set of four points (no three collinear) to a given set of four image points (no three collinear) is demonstrated in Example 4.4.

EXAMPLE 4.4. Find a matrix of the collineation that maps $P(1, -3, 2)$, $Q(2, -1, 3)$, $R(0, 3, -2)$, and $S(-1, 3, 0)$ to $P'(3, 7, 7)$, $Q'(0, 0, 1)$, $R'(5, 7, 6)$, and $S'(1, 9, 7)$, respectively.

The verification that no three of the points P, Q, R, and S are collinear requires the verification that none of the four determinants $|PQR|$, $|PQS|$, $|PRS|$, and $|QRS|$ are zero. Similar computations are required to show that no three of the points P', Q', R', and S' are collinear.

Following the procedure outlined in the proof of Theorem 4.34, we first find the matrix B by writing out the matrix equations for each of the equations: $s_1 P' = BX$, $s_2 Q' = BY$, $s_3 R' = BZ$, and $s_4 S' = BU$. The first of these equations becomes

$$\begin{bmatrix} b_{11} & b_{12} & b_{13} \\ b_{21} & b_{22} & b_{23} \\ b_{31} & b_{32} & b_{33} \end{bmatrix} \begin{bmatrix} 1 \\ 0 \\ 0 \end{bmatrix} = s_1 \begin{bmatrix} 3 \\ 7 \\ 7 \end{bmatrix} \quad \text{or} \quad \begin{aligned} b_{11} &= 3s_1 \\ b_{21} &= 7s_1 \\ b_{31} &= 7s_1 \end{aligned} \qquad (4.10)$$

Before writing out the matrix version of the second equation, we replace the first column by these values.

$$\begin{bmatrix} 3s_1 & b_{12} & b_{13} \\ 7s_1 & b_{22} & b_{23} \\ 7s_1 & b_{32} & b_{33} \end{bmatrix} \begin{bmatrix} 0 \\ 1 \\ 0 \end{bmatrix} = s_2 \begin{bmatrix} 0 \\ 0 \\ 1 \end{bmatrix} \quad \text{or} \quad \begin{aligned} b_{12} &= 0 \\ b_{22} &= 0 \\ b_{32} &= s_2 \end{aligned} \qquad (4.11)$$

By replacing the second column, the third equation becomes

$$\begin{bmatrix} 3s_1 & 0 & b_{13} \\ 7s_1 & 0 & b_{23} \\ 7s_1 & s_2 & b_{33} \end{bmatrix} \begin{bmatrix} 0 \\ 0 \\ 1 \end{bmatrix} = s_3 \begin{bmatrix} 5 \\ 7 \\ 6 \end{bmatrix} \quad \text{or} \quad \begin{aligned} b_{13} &= 5s_3 \\ b_{23} &= 7s_3 \\ b_{33} &= 6s_3 \end{aligned} \qquad (4.12)$$

Finally the fourth equation becomes

$$\begin{bmatrix} 3s_1 & 0 & 5s_3 \\ 7s_1 & 0 & 7s_3 \\ 7s_1 & s_2 & 6s_3 \end{bmatrix} \begin{bmatrix} 1 \\ 1 \\ 1 \end{bmatrix} = s_4 \begin{bmatrix} 1 \\ 9 \\ 7 \end{bmatrix} \quad \text{or} \quad \begin{aligned} 3s_1 \phantom{{}+s_2} + 5s_3 &= s_4 \\ 7s_1 \phantom{{}+s_2} + 7s_3 &= 9s_4 \\ 7s_1 + s_2 + 6s_3 &= 7s_4 \end{aligned} \qquad (4.13)$$

Applying straightforward row reduction to the coefficient matrix for these equations yields $s_1 = -19$, $s_2 = 24$, $s_3 = 10$, and $s_4 = -7$; so the matrix B can be obtained by substituting these values in the matrix in (4.13):

$$B = \begin{bmatrix} -57 & 0 & 50 \\ -133 & 0 & 70 \\ -133 & 24 & 60 \end{bmatrix}$$

Having found the matrix B, we now need to find the matrix C, which is determined by the four equations $s_5 P = CX$, $s_6 Q = CY$, $s_7 R = CZ$, and $s_8 S = CU$. We can simplify these calculations considerably by merely noting that we are again mapping the points X, Y, and Z with the matrix C so that the matrix equation comparable to (4.13) can be obtained by merely replacing the first, second, and third columns with $s_5 P$, $s_6 Q$, and $s_7 R$ giving:

$$\begin{bmatrix} 1s_5 & 2s_6 & 0 \\ -3s_5 & -1s_6 & 3s_7 \\ 2s_5 & 3s_6 & -2s_7 \end{bmatrix} \begin{bmatrix} 1 \\ 1 \\ 1 \end{bmatrix} = s_8 \begin{bmatrix} -1 \\ 3 \\ 0 \end{bmatrix} \quad \text{or} \quad \begin{array}{r} 1s_5 + 2s_6 = -s_8 \\ -3s_5 - s_6 + 3s_7 = 3s_8 \\ 2s_5 + 3s_6 - 2s_7 = 0 \end{array}$$

$$(4.14)$$

These equations yield $s_5 = 19$, $s_6 = -6$, $s_7 = 10$, and $s_8 = -7$; so the matrix C is given by

$$C = \begin{bmatrix} 19 & -12 & 0 \\ -57 & 6 & 30 \\ 38 & -18 & -20 \end{bmatrix}$$

Finally, computing the matrix product BC^{-1} gives the matrix of the collineation:

$$A = \begin{bmatrix} 2 & 1 & -1 \\ 0 & 3 & 1 \\ -1 & 2 & 0 \end{bmatrix}$$

Since collineations preserve collinearity, concurrence and cross ratios, Theorem 4.34 allows us to simplify analytic proofs involving these properties by choosing any four points, no three collinear, as the points X, Y, Z, and U, where as before we will assume these points have the following coordinates: $X(1,0,0)$, $Y(0,1,0)$, $Z(0,0,1)$, and $U(1,1,1)$. This technique is illustrated in the following proof that shows that our analytic model satisfies Axiom 4.

Verification of Axiom 4

Let $X(1,0,0)$, $Y(0,1,0)$, $Z(0,0,1)$, and $U(1,1,1)$ be the four points of a quadrangle. Using straightforward calculations, we can show that the diagonal points of this quadrangle are $XY \cdot UZ = A(1,1,0)$, $XZ \cdot UY = B(1,0,1)$, and $UX \cdot ZY = C(0,1,1)$. A quick computation shows that $|ABC| \neq 0$ so the diagonal points of the quadrangle are not collinear (see Exercise 7).

As indicated by Theorem 4.22, projectivities need not have any invariant elements. Collineations, on the other hand, always have at least one invariant point and one invariant line. The proof of this statement is a direct application of the theory of eigenvectors from linear algebra.

Theorem 4.35. *A collineation has at least one invariant point and one invariant line.*

Proof. To show that a collineation with matrix A has at least one invariant point, note that there will be an invariant point X iff there is a nonzero scalar s such that $sX = AX$. But $sX = AX$ iff $sIX - AX = (sI - A)X = 0$ where I is the identity matrix. This last equation has a nontrivial solution X iff $|sI - A| = 0$; but since A is a 3×3 matrix with real entries $|sI - A|$ is a third-degree polynomial in s and so has at least one real solution for s. (Note this solution cannot be 0.) To show that the collineation with matrix A has at least one invariant line, the same procedure is used, beginning with the equation $ku' = uA^{-1}$. □

The invariant line of a collineation need not be *pointwise* invariant; that is, even though points on the invariant line must remain on the line under the collineation, the points themselves may not remain fixed.

Definition 4.20. A collineation that has one pointwise invariant line is called a *perspective collineation*. The pointwise invariant line is called the *axis*.

By Theorem 4.34 a perspective collineation other than the identity can have at most one invariant point not on the axis. The following theorem demonstrates that there is always one invariant point under a perspective collineation.

Theorem 4.36. *Every perspective collineation has a linewise invariant point.* (*This point is called the* center).

Proof. Let m be the axis of the perspective collineation.
    Case 1. There is an invariant point not on m. Let this invariant point be called C. Then any line through C intersects m in a second invariant point

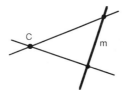

Figure 4.35

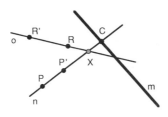

Figure 4.36

(Fig. 4.35). Thus each line through C has two invariant points and hence is invariant. It follows that C is linewise invariant.

Case 2. The only invariant points are those on m. Let P be any point not on m. Consider the line $n = PP'$ where P' is the image of P under the perspective collineation. Let $C = n \cdot m$. Then $n = CP = CP'$ is invariant. If R is another point not on m or n, there similarly exists an invariant line $o = RR'$. Let $X = o \cdot n$ (Fig. 4.36). Then since o and n are both invariant, it follows that X is invariant so X is on m. But $n \cdot m = C$, thus $X = C$. Therefore, every point not on m lies on an invariant line through C, or in other words, every line through C is invariant.   □

The proof of Theorem 4.36 shows that a perspective collineation with center C and axis m maps a given point P $(P \neq C)$ not on m to a point P' on PC. Subject to this condition, however, the image of P can be arbitrary. However, once the image of P is named, the image of any other point under a perspective collineation with a given axis and center is completely determined.

Theorem 4.37. *There exists a unique perspective collineation with axis m and center C that maps a given point P (P $\neq$ C and P not on m) to a given point P' on PC.*

Proof. *Case* 1. C is not on m. Let $PC \cdot m = D$ and let E and F be two additional points on m (Fig. 4.37). Then by Theorem 4.34 there exists a unique collineation that maps P to P', C to C, E to E, and F to F. Clearly, this collineation keeps m invariant since it keeps E and F invariant. Note that

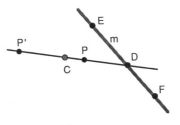

Figure 4.37

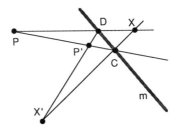

Figure 4.38

$PC = P'C$ is a second invariant line. Thus $D = PC \cdot m$ is a third invariant point on m, and it follows that the projectivity induced on m by this collineation is the identity (Theorem 4.8), and so m is pointwise invariant. Thus the collineation is a perspective collineation with axis m, and as in the proof of Theorem 4.36, the center can be shown to be C.

Case 2. C is on m. Let $PX \cdot m = D$ where X is a point not on either PC or m (Fig. 4.38). Then if a perspective collineation exists as desired, it must map X to $X' = CX \cdot P'D$ (see Exercise 10). But by Theorem 4.34, there exists a unique collineation mapping P to P', X to X', C to C, and D to D. As before, m is invariant under the collineation because C and D are invariant. However, m must be shown pointwise invariant and C linewise invariant. As shown in Problem 9, it is sufficient to show that C is linewise invariant and then note that since m has at least two invariant points, it must be the axis.

To show that C is linewise invariant, note that $CP = CP'$, m and $CX = CX'$ are three invariant lines through C. Thus by the dual of the argument in case 1, C is linewise invariant and the result follows.   □

These collineations are called perspective collineations since they map triangles to perspective triangles. The proof of this result follows directly from the definitions of the center and axis of a perspective collineation and from the definition of perspective triangles.

Theorem 4.38. $\triangle P'Q'R'$ *is the image of* $\triangle PQR$ *under a perspective colline-ation with center* C *and axis* m, *iff the triangles are perspective from the point* C *and perspective from the line* m.

Proof. (a) For the first half of the proof, see Exercise 11.

(b) Now assume that $\triangle PQR$ and $\triangle P'Q'R'$ are perspective from C and m. Since $\triangle PQR$ is a triangle, the three points P, Q, R are not collinear. Thus at least one of the points, say P, is not on m. Furthermore since $\triangle PQR$ and $\triangle P'Q'R'$ are perspective from C, P' is on PC. So by Theorem 4.37, there is a perspective collineation T with center C and axis m that maps P to P'. It must now be shown that $T(Q) = Q'$ and $T(R) = R'$.

By the proof of Theorem 4.37 $T(Q) = P'D \cdot QC$ where $D = PQ \cdot m$. But

$PQ \cdot m = P'Q' \cdot m$ since the triangles are perspective from m, so $P'D = P'Q'$. Thus $T(Q) = P'Q' \cdot QC$. But Q' is on QC since the triangles are perspective from C, so $T(Q) = Q'$. Likewise $T(R) = R'$.

As the proof of Theorem 4.37 indicates, there is a distinction between the perspective collineations that have their centers on their axis and those that do not.

Definition 4.21. A perspective collineation other than the identity is called an *elation* if its center lies on its axis and a *homology* if its center does not lie on its axis.

Homologies have another property worthy of note.

Theorem 4.39. *Under a homology with center C and axis m, any point P not on m ($P \neq C$) has an image P' such that C, P, and P' are collinear and if $m \cdot CP = Q$, then $R(C, Q, P, P')$ is constant for all P.*

Proof. The fact that C, P, and P' are collinear for all perspective collineations has been noted previously.

Case 1. X is a point not on CP or on m. Let X' be its image under the homology, and let $D = CX \cdot m$, $E = PX \cdot m$. So $X' = XC \cdot EP'$ (Fig. 4.39). Then $CQPP' \overset{E}{\wedge} CDXX'$, and thus by Theorem 4.26, $R(C, Q, P, P') = R(C, D, X, X')$.

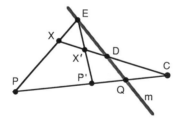

Figure 4.39

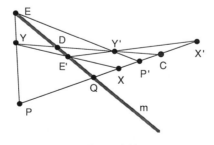

Figure 4.40

Case 2. X is a point on CP. Then let Y be a point not on CP (Fig. 4.40). By case 1, $R(C, Q, P, P') = R(C, D, Y, Y')$ and Y can then be used in place of P in case 1 to yield a similar result for X. □

If this constant cross ratio is -1, the homology is called a *harmonic homology*.

EXERCISES

1. Prove that the set of points of the analytic model (i.e., nonzero equivalence classes of ordered triples from R^3), together with the equivalence class $\{(0,0,0)\}$, form a vector space under the usual addition and scalar multiplication in R^3.

2. Prove Theorem 4.30.

3. Given the collineation with matrix

$$\begin{bmatrix} 2 & 0 & 1 \\ 0 & 2 & 3 \\ 0 & 0 & 1 \end{bmatrix}$$

(a) Write out the point equation for this collineation and find P' and Q', the images of $P(1, 2, 3)$ and $Q(-1, 0, 1)$, respectively. (b) Find the coordinates of the line $P'Q'$. (c) Write out the line equation for this collineation and find the image of $l[1, -2, 1]$. (d) Do your answers for parts *b* and *c* agree?

4. Show that the relation " $\sim$ " defined here is an equivalence relation on the set of 3×3 matrices: $A \sim B$ iff $A = kB$ for some nonzero scalar k.

5. Prove Theorem 4.32.

6. Find the matrix of the collineation that maps $P(1, 0, 1)$, $Q(2, 0, 1)$, $R(0, 1, 1)$, and $S(0, 2, 1)$ to $X(1, 0, 0)$, $Y(0, 1, 0)$, $Z(0, 0, 1)$, and $U(1, 1, 1)$, respectively.

7. Complete the details in the verification of Axiom 4.

8. Find the invariant points and lines of the collineations with the following matrices:

(a) $\begin{bmatrix} 1 & 1 & 0 \\ 0 & 1 & 0 \\ 0 & 0 & a \end{bmatrix}$ (b) $\begin{bmatrix} 1 & 0 & 0 \\ 0 & 1 & 0 \\ 0 & 1 & 1 \end{bmatrix}$.   (c) $\begin{bmatrix} a & 1 & 0 \\ 0 & a & 1 \\ 0 & 0 & a \end{bmatrix}$

where $a \neq 0, 1$.   where $a \neq 0$.

9. Prove: Every collineation with a linewise invariant point is a perspective collineation.

10. Show that in the proof of case 2 of Theorem 4.37, $X' = CX \cdot P'D$.

11. Prove the first part of Theorem 4.38.

12. Show that the collineation with the following matrix is a homology. Is it a harmonic homology?

$$\begin{bmatrix} 1 & 0 & 0 \\ 0 & 1 & 0 \\ 2 & 0 & -1 \end{bmatrix}$$

13. Find the matrix of an elation with axis $[0,0,1]$ and center $(1,0,0)$.

4.11. Correlations and Polarities

The second type of transformations of the projective plane, known as correlations, are also one-to-one linear transformations. Here, however, the images of points are lines.

Definition 4.22. A *correlation* is a one-to-one linear transformation of the set of points of the projective plane onto the set of lines of the projective plane.

Correlations, too, can be represented by 3×3 matrices with matrix equations much like those used for collineations; except in this case, the ordered triples resulting from these mappings are interpreted as homogeneous coordinates of lines. Just as for collineations, there is an entire equivalence class of matrices corresponding to each correlation. These and several other results characterizing properties of correlations can be proved using arguments nearly identical to those used to prove similar results about collineations.

Theorem 4.40. *A correlation can be represented by a 3×3 real valued matrix A where $|A| \neq 0$. The matrix equation for the correlation is $su^t = AX$ where $X \in R^3$ and $s \neq 0$.*

Theorem 4.41. *A correlation maps collinear points to concurrent lines. The image of a line u under a correlation with matrix A is given by the equation $kX^t = uA^{-1}, k \neq 0$.*

Corollary. *Under a correlation, concurrent lines are mapped to collinear points.*

Theorem 4.42. *A correlation of the projective plane induces a projectivity between the elements of the corresponding pencils.*

Corollary. *Cross ratios and harmonic sets are invariant under correlations.*

Theorem 4.43. *There exists a unique correlation that maps any four points, no three collinear, to any four lines, no three concurrent.*

Thus a given correlation that maps points to lines according to the equation $su^t = AX$ also maps lines to points according to the equation $kX^t = uA^{-1}$. (The transpose is used in both equations, since points are represented by column matrices and lines are represented by row matrices.) In general, correlations map any given set to the dual set. For example, the image of a quadrangle under a correlation is a quadrilateral, and vice versa. It follows that correlations give us an analytic method for studying duality.

Since correlations map points to lines and lines to points, it seems reasonable to expect that a correlation that maps a point P to a line p will automatically map the line p to the point P. However, this does not necessarily happen, since a correlation that maps a point X to a line u according to $su^t = AX$ will map the line u to a point Y according to $kY^t = uA^{-1}$. Solving the first equation for u gives $u = (1/s)(AX)^t = (1/s)X^tA^t$. If each point X mapped to line u, which in turn mapped back to X, so that $X = Y$ for each point X, then $kX^t = k'Y^t = uA^{-1} = ((1/s)X^tA^t)A^{-1}$ or $skX^t = X^t(A^tA^{-1})$. This will hold for all possible points X iff $A^tA^{-1} = I$, that is, iff $A^t = A$. So a correlation maps every point X to a line u and the line u back to X iff its matrix is symmetric. Correlations of this type are called *polarities*. Since we shall soon show that the set of polarities give analytic expression to conics, we begin using the letter C to denote matrices of polarities.

Definition 4.23. A correlation whose matrix is symmetric is called a *polarity*. If a polarity maps a point P to a line p (and thus p to P), then p is called the *polar* of P and P is called the *pole* of p with respect to the given polarity.

Since polarities are correlations, they are one-to-one mappings; hence polars of distinct points are distinct lines, and vice versa. This polarity relation also has the characteristic property first described in Section 1.5 in that it pairs points that lie on each others' polars. Such points are called *conjugate* points with respect to the polarity.

Theorem 4.44. *A point P is on the polar of a point Q under a given polarity iff Q is on the polar of P under this same polarity.*

Proof. Let C be the matrix of the polarity and let q and p be the polars of Q and P, that is, $s_1q^t = CQ$ and $s_2p^t = CP$. Since P is on the polar of Q, $qP = 0$; but $s_1q = Q^tC$, so $Q^tCP = 0$. Transposing gives $P^tCQ = 0$ or $pQ = 0$, that is, Q is on the polar of P.   □

Corollary. *P is on the polar of Q with respect to a polarity with matrix C iff $Q^tCP = 0$. And p contains the pole of q with respect to this same polarity iff $pC^{-1}q^t = 0$.*

Definition 4.24. Two points are called *conjugate points* with respect to a given polarity if each point is on the polar of the other. A point that lies on its own polar is called a *self-conjugate point* with respect to the given polarity.

Two lines are called *conjugate lines* with respect to a given polarity if each line is incident with the pole of the other. A line that is incident with its own pole is called a *self-conjugate line* with respect to the given polarity.

The corollary to the previous theorem leads directly to a matrix equation for sets of self-conjugate points (see Exercise 2). Multiplying out the matrix product in this equation yields a quadratic form whose similarity to the quadratic forms encountered in Section 3.9 should be suggestive of a connection between sets of self-conjugate points and point conics.

Theorem 4.45. *The set of self-conjugate points of a polarity with matrix C is the set of points X satisfying the equation $X^tCX = 0$. The set of self-conjugate lines of this same polarity is the set of lines satisfying the equation $uC^{-1}u^t = 0$.*

Corollary. *The set of self-conjugate points of a polarity with matrix C is the set of points X satisfying the equation:*

$$c_{11}x_1^2 + c_{22}x_2^2 + c_{33}x_3^2 + 2c_{12}x_1x_2 + 2c_{13}x_1x_3 + 2c_{23}x_2x_3 = 0$$

Using the matrix equation for a set of self-conjugate points, we can now show that such sets are preserved under collineations. A similar procedure can be used to show that the pole–polar relation is also preserved under these mappings, that is, if P and p are pole and polar with respect to a polarity with matrix C; then P' and p', their images under a collineation, will be pole and polar with respect to the polarity with matrix C' (see Exercise 3).

Theorem 4.46. *A collineation with matrix A maps a set of self-conjugate points with matrix C to a set of self-conjugate points with matrix $C' = (A^{-1})^tC(A^{-1})$.*

Proof. Let S be a set of self-conjugate points with equation $X^tCX = 0$ where C is a 3×3 nonsingular, symmetric matrix. Let A be the matrix of an arbitrary collineation. Then A is also a 3×3 nonsingular matrix, and the corresponding point equation is $sX' = AX$. Solving for X and X^t gives $X = sA^{-1}X'$ and $X^t = s(X')^t(A^{-1})^t$. Substituting into the equation $X^tCX = 0$ yields $(X')^t(A^{-1})^tC(A^{-1})X' = 0$ or $(X')^t((A^{-1})^tCA^{-1})X' = 0$. But $(A^{-1})^tCA^{-1}$ is a 3×3 nonsingular, symmetric matrix and hence the matrix of a polarity. Thus X is in the set S of self-conjugate points with matrix C iff X' is in the set S' of self-conjugate points with matrix $C' = (A^{-1})^tCA^{-1}$.   □

The previous theorem will allow us to simplify our work with self-conjugate sets of points by "assigning" special coordinates to some of the points involved, much as we did when we verified Axiom 4 in Section 4.10. In particular, we use this technique to demonstrate that sets of self-conjugate points, which were defined analytically in terms of polarities, are point conics, figures that can be constructed entirely with points and lines. Even with this simplifying technique, the proof of this result is somewhat long and involved, but it nicely illustrates the use of analytic methods in projective geometry. The significance of the theorem and corollaries make the effort worthwhile.

Theorem 4.47. *A nonempty set of self-conjugate points with respect to a given polarity is a point conic and a nonempty set of self-conjugate lines with respect to a given polarity is a line conic. Conversely, any point conic is a set of self-conjugate points with respect to some polarity and any line conic is a set of self-conjugate lines with respect to some polarity.*

Proof. By the principle of duality, it is sufficient to verify the result for point conics.

Let $\mathscr{C}$ be a nonempty set of self-conjugate points. Since $\mathscr{C}$ is nonempty, we can show that $\mathscr{C}$ contains at least three distinct, noncollinear points (see Exercise 6). We will assume these points are $X(1,0,0)$, $Z(0,0,1)$, and $U(1,1,1)$ and that the polars at X and Z intersect at $Y(0,1,0)$. Since X and Z are self-conjugate points, their polars relative to $\mathscr{C}$ are $XY[0,0,1]$ and $ZY[1,0,0]$. Algebraically, this means we need a symmetric matrix C that satisfies the following:

$$C[1,0,0]^t = s_1[0,0,1]^t \quad \text{and} \quad C[0,0,1]^t = s_2[1,0,0]^t$$

These equations yield $c_{11} = c_{12} = c_{23} = c_{33} = 0$ and $c_{13} \neq 0$. Finally, requiring that the point U also be self-conjugate gives $c_{22} = 1$ and $c_{13} = -\frac{1}{2}$ so the equation of $\mathscr{C}$ becomes $(x_2)^2 - x_1 x_3 = 0$. It is then sufficient to show that the set of points satisfying this equation is point conic, that is, a set of points of intersection of corresponding lines of two projectively related pencils of lines.

Let X and Z be centers of two pencils. The projectivity we will use is uniquely determined by the correspondence: $XY\,XZ\,XU \wedge ZX\,ZY\,ZU$. Note that under this projectivity, X, Z, and U are all points of intersection of corresponding lines; and since XY corresponds to ZX, the line between the centers of the two pencils, it will be a tangent at X. Similarly, ZY will be a tangent at Z. Letting XY and XZ be base lines of the first pencil and ZX and ZY base lines of the second pencil, will give us a projectivity with a diagonal matrix (see Exercise 1, Section 4.8). Finally, requiring that $XU[0, 1, -1]$ with homogeneous parameters $(-1, 1)$ map to $ZU[1, -1, 0]$, also with homogeneous parameters $(-1, 1)$, gives the 2×2 identity matrix as the matrix of the projectivity.

To show that $\mathscr{C}$ is exactly the set of points of intersection of corresponding lines under this projectivity, let $P(p_1, p_2, p_3)$ be an arbitrary point of $\mathscr{C}$. Then the projectivity will map the line $XP = l[0, -p_3, p_2]$ with homogeneous parameters $(p_2, -p_3)$ to line l' through Z with the same homogeneous parameters. So l' has coordinates $[-p_3, p_2, 0]$. Using the determinant condition to find the point $l \cdot l'$ gives $(-(p_2)^2, -p_3 p_2, -(p_3)^2)$ as the coordinates of this point of intersection, but since P is a point $\mathscr{C}$, $(p_2)^2 = p_1 p_3$ so the point $l \cdot l'$ has coordinates (p_1, p_2, p_3). In other words, the point P is the point of intersection of the projectively related lines l and l' iff P is in $\mathscr{C}$.

To complete the first half of the proof, we need to verify that this projectivity is not a perspectivity. To do this, it is sufficient to note that the line XZ, which joins the two centers of the pencils, does not correspond to itself.

Conversely, to show that any point conic $\mathscr{C}$ is the set of self-conjugate points with respect to a polarity we can use a similar procedure. Let P, Q, R be three distinct points of $\mathscr{C}$ and let S be the point of intersection of the tangents to $\mathscr{C}$ at P and Q. Then P, Q, R, and S are four distinct points, no three collinear (see Exercise 7). Since collineations preserve incidence and therefore conics, we can assume that P, Q, R, and S are the points $X(1, 0, 0)$, $Z(0, 0, 1)$, $U(1, 1, 1)$, and $Y(0, 1, 0)$, respectively.

By the corollary to Theorem 4.15, the tangents at X and Z, together with the three points X, Z, and U, uniquely determine the conic, so it is sufficient to show that these tangents and points determine a polarity with matrix C relative to which $\mathscr{C}$ is a set of self-conjugate points. Since in the first part of the proof, the two self-conjugate lines became tangents to the conic, here we will find a polarity under which the two tangent lines are self-conjugate. This, together with the condition that U be a self-conjugate point, leads to the same equation as before, namely, $(x_2)^2 - x_1 x_3 = 0$. So there is indeed a polarity under which $\mathscr{C}$ is a set of self-conjugate points. $\square$

Corollary 1. *A point conic has an equation of the form $X^t C X = 0$ and a line conic has an equation of the form $u C^{-1} u^t = 0$ where C is a symmetric, nonsingular 3×3 matrix.*

Therefore, any point conic corresponds to a symmetric matrix that is the matrix of a polarity. This polarity matrix is called the *matrix of the point conic*. Furthermore, if line p corresponds to point P under the polarity determined by the conic, P and p are said to be pole and polar *with respect to the conic*. This terminology is used in the statement of two more corollaries to Theorem 4.47, which formalize the relationship between self-conjugate lines and tangents.

Corollary 2. *Let P be a point of a point conic $\mathscr{C}$. The polar of P with respect to $\mathscr{C}$ is the tangent at P; conversely, the tangent to $\mathscr{C}$ at P is the polar of P with respect to $\mathscr{C}$.*

Corollary 3. *If X is a point of a point conic $\mathscr{C}$, with matrix C, then u, the tangent to $\mathscr{C}$ at X, is given by the equation $su^t = CX$.*

Using this last corollary, we can show that the line conic determined by a given polarity consists of the tangents to the point conic determined by the same polarity.

Theorem 4.48. *The tangents to a point conic are the lines of the line conic determined by the same polarity.*

Proof. Let X be a point on a point conic with matrix C. By Corollary 3 to

Theorem 4.47, u, the tangent at X, is given by $su^t = CX$. Solving this equation for X, gives $X = sC^{-1}u^t$.

Since X is on the point conic, $X^tCX = 0$ by Corollary 1 of the same theorem. Substituting the previous expression for X into this equation gives $(sC^{-1}u^t)^tC(sC^{-1}u^t) = 0$, or $uC^{-1}CC^{-1}u^t = uC^{-1}u^t = 0$. So u satisfies the equation of the line conic determined by the same polarity.   □

A polarity also determines polars of points not on the corresponding conic. The following theorems and definitions yield a method of constructing polars of other points (and, by duality, poles of lines other than tangents). These constructions will assume added importance in Section 4.12 when we describe non-Euclidean geometries as subgeometries of projective geometry.

Theorem 4.49. *The point of intersection of two tangents to a point conic is the pole of the line joining the points of tangency.*

Proof. Let p and q be tangents to a point conic at points P and Q, respectively; that is, p and q are the polars of P and Q, respectively. Let $R = p \cdot q$ (Fig. 4.41). Then R is on both the polar of P and the polar of Q so by Theorem 4.44, P and Q are both on the polar of R, so PQ is the polar of R, and therefore R is the pole of line PQ by definition.   □

Corollary. *Any point lies on at most two tangents to a given point conic.*

The proof of Theorem 4.49 demonstrates the existence of triangles where one vertex is the pole of the opposite side. Triangles in which each vertex is the pole of the opposite side are particularly significant. That triangles like this do exist is demonstrated by the next theorem.

Definition 4.25. If each vertex of a triangle is the pole of the opposite side of the triangle with respect to a conic, then the triangle is said to be *self-polar* relative to the conic.

Theorem 4.50. *If A, B, C, and D are four distinct points of a point conic then the diagonal triangle of quadrangle ABCD is self-polar.*

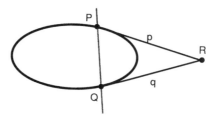

Figure 4.41

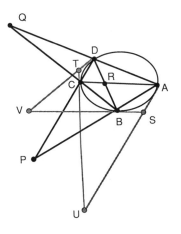

Figure 4.42

Proof. Let $P = CD \cdot AB$, $Q = CB \cdot AD$, and $R = AC \cdot BD$ be the diagonal points of the quadrangle, and let

$$S = \tan B \cdot \tan A \quad \text{and} \quad T = \tan C \cdot \tan D$$
$$U = \tan C \cdot \tan A \quad \text{and} \quad V = \tan D \cdot \tan B$$

Then by a corollary to Theorem 4.14, Q, R, S, and T are collinear as are P, Q, U, and V (Fig. 4.42). By Theorem 4.49, P is on the polars of both S and T, so $TS = QR$ is the polar of P. Similarly, R is on the polars of both U and V, so $UV = PQ$ is the polar of R. And finally, since Q is on the polars of P and R, then PR is the polar of Q.

Corollary. *If a line m through a point P not on a point conic intersects the conic, the points of intersection are harmonic conjugates with respect to P and the point of intersection of m with the polar of P.*

Theorem 4.49 indicates how to construct poles of lines that intersect a conic twice. Lines that intersect a conic exactly once are just the tangents and hence the polars of the point of intersection, but there are lines that do not intersect the conic at any point. This distinction provides the basis for the next definition.

Definition 4.26. If the polar of P with respect to a given conic does not intersect a given point conic, P is said to be an *interior point* of the conic. If the polar of P with respect to a given point conic intersects the conic in two distinct points, P is said to be an *exterior point* of the conic

In order to demonstrate the construction of polars of interior and exterior points and the construction of poles of lines that do and do not contain any interior points, we will make use of the following lemma (see Exercise 11).

Lemma. *A line contains interior points of a point conic iff it intersects the conic at two distinct points.*

Construction of Poles and Polars

Case 1. *Construction of the Polar of a Point P not on the Conic.* Let l and m be two lines through P, both of which intersect the conic $\mathscr{C}$ at two points. Let A and B be the points of intersection of l with $\mathscr{C}$ and let C an D be the points of intersection of m with $\mathscr{C}$. Then A, B, C, and D form a quadrangle so by Theorem 4.50, its diagonal triangle is self-polar. In other words, the line joining $Q = AC \cdot BD$ and $R = AD \cdot BC$ is the polar of P (see Fig. 4.43).

Case 2. *Construction of the Pole of a Line p Not Tangent to a Conic.* If p intersects the conic at distinct points R and S, then $P = \tan R \cdot \tan S = $ pole of p by Theorem 4.49 (Fig. 4.44). If p does not intersect the conic, let R and S be two distinct points on p. Then since p does not intersect the conic, it follows from the preceding lemma that all points on p and in particular R and S, are exterior points of the conic. Hence their polars r and s, respectively, each intersect the conic twice. Let $P = r \cdot s$ (Fig. 4.45). Then, since P is on the polar of R and the polar of S, it follows that P is the pole of p. (Note that P is an interior point.)

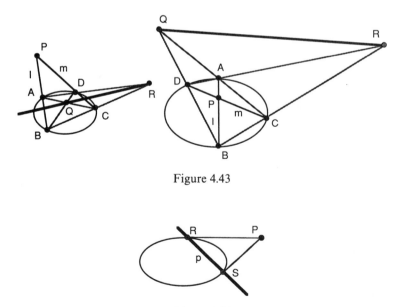

Figure 4.43

Figure 4.44

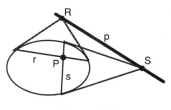

Figure 4.45

Self-polar triangles are also used in mapping a particular point conic to another point conic in *standard form*. (A conic in standard form will play an important role in the next section.) The proof of the first of the theorems necessary to achieve this mapping involves a special self-polar triangle and makes use of techniques similar to those used in part of the proof of Theorem 4.47 (see Exercise 13).

Theorem 4.51. *Triangle* $\triangle XYZ$ *[where* $X(1,0,0)$, $Y(0,1,0)$, *and* $Z(0,0,1)$*] is a self-polar triangle with respect to a conic iff the matrix of the conic is diagonal.*

Thus any point conic is *equivalent*, that is, can be mapped via a collineation, to a conic with an equation of the form $a(x_1)^2 + b(x_2)^2 + c(x_3)^2 = 0$. However, the next theorem shows that conics with equations of this form can, in turn, be mapped to conics with yet a simpler equation. This theorem will even include the case where the original point conic contains no points in the real projective plane.

Theorem 4.52. *Any point conic is projectively equivalent to a conic with an equation of the form* $(x_1)^2 + (x_2)^2 \pm (x_3)^2 = 0$ *(i.e., any conic can be mapped via a collineation to a conic with this equation).*

Proof. Let $\triangle PQR$ be a self-polar triangle with respect to a given point conic $\mathscr{C}$, and let T be a collineation that maps P, Q, R, to X, Y, Z, respectively. Then $\triangle XYZ$ will be self-polar with respect to the conic $T(\mathscr{C})$, so this latter conic will have a diagonal matrix by Theorem 4.51. Now either all of the diagonal entries are of the same sign or one of the entries differs in sign from the other two. In the first case, we will use a matrix representation in which the diagonal entries are all positive. In the second case we can, if necessary, make use of a collineation that switches an appropriate pair of the points X, Y, and Z to obtain a conic with a diagonal matrix representation with a negative third entry (see Exercise 15). So we can assume that the matrix representation of the image conic is of the following form where a and b are positive and c is nonzero:

$$C = \begin{bmatrix} a & 0 & 0 \\ 0 & b & 0 \\ 0 & 0 & c \end{bmatrix}$$

Finally, let S be the collineation with matrix

$$A = \begin{bmatrix} \sqrt{a} & 0 & 0 \\ 0 & \sqrt{b} & 0 \\ 0 & 0 & \sqrt{|c|} \end{bmatrix}$$

Then by Theorem 4.46 the conic $ST(\mathscr{C})$ will have matrix $C' = (A^{-1})C(A^{-1})$. Computing this matrix product gives

$$C' = \begin{bmatrix} 1 & 0 & 0 \\ 0 & 1 & 0 \\ 0 & 0 & \pm 1 \end{bmatrix}$$

So the conic $ST(\mathscr{C})$ has an equation of the form $(x_1)^2 + (x_2)^2 \pm (x_3)^2 = 0$. $\qquad\square$

A point conic whose equation is of this form is said to be in *standard form*. The two possible standard forms determine two types of polarities. The names assigned to these two types are meant to be suggestive of two non-Euclidean geometries. In the next section, we explore the connection between these polarities and the corresponding geometries.

Definition 4.27. A polarity whose associated conic is equivalent to the conic with equation $(x_1)^2 + (x_2)^2 - (x_3)^2 = 0$ is called *hyperbolic*. A polarity whose associated conic is equivalent to the conic with equation $(x_1)^2 + (x_2)^2 + (x_3)^2 = 0$ is called *elliptic*.

EXERCISES

1. Given the polarity with matrix

$$\begin{bmatrix} 2 & 0 & -1 \\ 0 & 1 & 1 \\ -1 & 1 & 0 \end{bmatrix}$$

(a) Find the equations of the sets of self-conjugate points and self-conjugate lines determined by this polarity. (b) Find the pole of the line $[1, 1, 1]$. (c) Find a point conjugate to the point $(1, 1, 1)$.

2. Prove Theorem 4.45 and its corollary.

3. Show that the pole–polar relation is preserved under a collineation.

In Exercises 4–6, $\mathscr{C}$ is a nonempty self-conjugate set of points determined by a given polarity.

4. Using Theorem 4.44, prove: If P is a point of $\mathscr{C}$, then the polar of P contains exactly one point of $\mathscr{C}$.

5. Prove: If A is a point of $\mathscr{C}$ and $B \neq A$ is a second point on the polar of A,

then the polar of B contains exactly two points of $\mathscr{C}$. [*Hint*: Assume A and B are the points $Z(0,0,1)$ and $Y(0,1,0)$, respectively.]

6. Use the result of Exercise 5 to show that $\mathscr{C}$ contains at least three noncollinear points.

7. Show that the four points P, Q, R, and S chosen in the proof of the second half of Theorem 4.47 are distinct with no three collinear.

8. Given the conic $(x_1)^2 + 2(x_2)^2 + 5(x_3)^2 - 2x_2x_3 - 2x_1x_3 - 4x_1x_2 = 0$ find (a) the tangent at point $(1,1,1)$, (b) the polar of $(3,1,5)$, and (c) the tangents from the point $(1, -2, 0)$.

9. Prove the corollary to Theorem 4.49.

10. Prove the corollary to Theorem 4.50. [*Hint*: Let n be a second line through P intersecting the conic at two points. Find a harmonic set formed by these four points and then use a perspectivity.]

11. Let $\mathscr{C}$ be the conic with equation $(x_2)^2 - x_1x_3 = 0$. (a) Show that the point $P(p_1, p_2, p_3)$ is an interior point of $\mathscr{C}$ iff $(p_2)^2 - p_1p_3 < 0$. (In general, P is an interior point of the conic $X^tCX = 0$ iff $P^tCP < 0$) (b) Show that every line contains exterior points of $\mathscr{C}$. (c) Use parts (a) and (b) to show that any line contains interior points of $\mathscr{C}$ iff it intersects $\mathscr{C}$ at two distinct points.

12. In the construction of the polar of a point P not on a conic, l and m were chosen as two lines through P, both of which intersected the conic twice. Describe how the polar of P could be obtained if l and/or m intersected the conic exactly once. Will this happen if P is an interior point?

13. Prove Theorem 4.51.

14. Show that a harmonic homology whose center and axis are pole and polar with respect to a point conic $\mathscr{C}$ keeps $\mathscr{C}$ invariant (i.e., it maps points on $\mathscr{C}$ back to points on $\mathscr{C}$).

15. If $\mathscr{C}$ is the point conic with equation $a(x_1)^2 + b(x_2)^2 + c(x_3)^2 = 0$ where $a < 0, b > 0, c > 0$, find the matrix of a collineation T such that the point conic $T(\mathscr{C})$ has an equation $a'(x_1)^2 + b'(x_2)^2 + c'(x_3)^2 = 0$ where $a' > 0, b' > 0$, and $c' < 0$. [*Hint*: Use a collineation that interchanges the points X and Z and keeps Y invariant.]

16. Show that the point conic determined by an elliptic polarity contains no points in the real projective plane.

17. Show that any collineation with an orthogonal matrix A will keep the conic determined by an elliptic polarity invariant. [*Hint*: Use the standard form of the conic and note that the matrix A is *orthogonal* iff $A^t = A^{-1}$.]

18. Prove: If T is a correlation then T^2 is a collineation. If A is the matrix of T, what is the matrix of T^2?

4.12. Subgeometries of Projective Geometry

In this final section, we see that Klein's definition of geometry allows us to view projective geometry as an *umbrella* geometry under which affine, similarity, Euclidean, hyperbolic, and single elliptic geometries all reside. Our approach is to demonstrate that the respective plane geometries can all be obtained as subgeometries of plane projective geometry. Although we do not do so, this approach can be extended to demonstrate a similar relation among the corresponding three-dimensional geometries.

To obtain the appropriate viewpoint, we begin by selecting an *absolute polarity* (its corresponding conic is known as the *absolute conic* $\mathscr{C}$) for each of the three geometries and then demonstrate that fundamental concepts of each geometry can be defined in terms of properties left invariant under a group of transformations that preserve $\mathscr{C}$. Since we have already explored these geometries in some depth in Chapters 2 and 3, we do not dwell on the details of each geometry here. Rather, we concentrate on identifying concepts in terms of their projective counterparts and indicate ways in which theorems of projective geometry can be used to verify standard results in each geometry. The ease with which we are able to make these identifications and prove these results should increase your apreciation for both the significance and beauty of projective geometry.

As indicated earlier, we begin by considering a particular polarity of the projective plane. We refer to this polarity and its associated point conic as the *absolute polarity* and *absolute conic*. As we discovered in Section 4.11, this absolute conic can be represented in standard form as $(x_1)^2 + (x_2)^2 \pm (x_3)^2 = 0$. In order to obtain both non-Euclidean and affine geometries, and eventually Euclidean geometry as subgeometries, we rewrite this equation as $c[(x_1)^2 + (x_2)^2] + (x_3)^2 = 0$, where $c = \pm 1$ or 0. The first two values of c yield the standard form given earlier, whereas $c = 0$ yields the degenerate conic $x_3 = 0$, that is, the line with coordinates $[0, 0, 1]$. So in this last case, the absolute conic consists of the ideal points, which were added to the affine plane to obtain an analytic model of the projective plane (see Section 4.7). It should then come as no surprise that the geometry obtained using this absolute conic is affine geometry. In the cases where $c = \pm 1$, we will also refer to points of the absolute conic as *ideal* points. The polarities determining the absolute conic when $c = 1$ and $c = -1$ are called elliptic and hyperbolic (Definition 4.27), since we shall see that these polarities determine elliptic and hyperbolic geometries, respectively.

We use the following procedure to demonstrate the relationship of each of these geometries to projective geometry. For each value of c, we select an absolute polarity and corresponding absolute conic $\mathscr{C}$. Using $\mathscr{C}$, we describe a special subset of points of the real projective plane. This subset of (*ordinary*) *points* will be the points remaining after the ideal (and in the case of hyperbolic geometry, ultraideal) points are deleted. In the case where $c = 1$, there are no *real* points on the absolute conic, so no points are deleted; that is, the set of points of the elliptic plane is identical to the set of points of the projective plane. (However, the absolute polarity that determines this *empty* conic is still useful.) It is important to observe that for each subgeometry, the only points of the geometry will be the (ordinary) points; that is, the ideal and ultraideal points are *not* points of the geometry. However, in order to make use of concepts in projective geometry, we will consider the point set of each geometry as a subset of the point set of the projective plane. We refer to this process as *embedding* the geometry into the projective plane. This embedding allows us to use the ideal and ultraideal points in addition to the (ordinary) points.

Following the identification of the appropriate set of points, we will list several definitions indicating how basic concepts of each geometry can be defined via properties of the projective plane. As you should note, these definitions are all stated in terms of projective properties that remain invariant under collineations that preserve the absolute conic. After we describe the definitions used in the non-Euclidean geometries, our presentation will focus on affine geometry; since this geometry includes the most familiar geometry, Euclidean. This approach will allow us to see that, with a slight stretch of the definition of a line conic, the cross ratio can be used to give a common definition of angle measure in all three geometries. Even though a similar definition can be used to define distance in the two non-Euclidean geometries, it is not possible to extend this definition to affine geometry.

Stating the definitions in terms of projective properties enables us to use theorems of projective geometry to verify that the concepts so defined do indeed have other expected properties. However, since the purpose of this section is to merely illustrate the relationships between the geometries, we will only indicate how a few such properties can be verified. You will be asked to verify a number of other properties. These exercises should increase your appreciation for the interrelatedness of the geometries we have studied, as well as provide an opportunity to review ideas from this chapter.

Hyperbolic Geometry

Ideal Points: Points of the absolute conic $\mathscr{C}$: $(x_1)^2 + (x_2)^2 - (x_3)^2 = 0$.
Ultraideal Points: Points exterior to $\mathscr{C}$.
(Ordinary) Points: Points of the real projective plane interior to $\mathscr{C}$.
Lines: Open chords of $\mathscr{C}$ (i.e., parts of projective lines containing points interior to $\mathscr{C}$).

Hyperbolic Definitions

1h: Two hyperbolic lines are *sensed parallel* if the corresponding projective lines intersect in ideal points.

2h: Two hyperbolic lines are *ultraparallel* if the corresponding projective lines intersect in ultraideal points.

3h: Two hyperbolic lines are *perpendicular* if the corresponding projective lines are conjugate with respect to the absolute conic.

4h: If A and B are two hyperbolic points, the *directed hyperbolic distance*, $d_h(A, B) = k \ln (R(A, B, P, Q))$ where P and Q are ideal points of line AB, and "ln" represents the natural logarithm.

5h: If a and b are intersecting hyperbolic lines, the *hyperbolic angle measure*, $m_h(\angle (a, b)) = k' \ln (R(a, b, p, q))$ where p and q are tangents to $\mathscr{C}$ from the point $a \cdot b$.

Using these definitions, we can *embed* the hyperbolic plane into the projective plane and use theorems of projective geometry to construct proofs of hyperbolic theorems. We will illustrate this process by verifying an extension of Theorem 53h (see Section 2.8). Note that hyperbolic and projective lines are not identical, since hyperbolic lines contain only those points of the corresponding projective line that are interior to the absolute conic $\mathscr{C}$. To keep track of this distinction, we will denote by l' the projective line that corresponds to the hyperbolic line l. We will also find it helpful to make use of a diagram within the projective plane. This diagram should look familiar, since it is merely a depiction within the Klein model described in Section 2.3.

Property 1h. Two hyperbolic lines are ultraparallel iff they have a unique common perpendicular.

Proof. (a) Let l and m be two ultraparallel lines. Then the corresponding projective lines l' and m' intersect in a point P' where P' is exterior to $\mathscr{C}$ (see Fig. 4.46). Let p' be the polar of P'. Since P' is exterior to $\mathscr{C}$, p' intersects $\mathscr{C}$ at two distinct points and therefore determines a hyperbolic line p (see the lemma in Section 4.11). Furthermore, p' is conjugate to both l' and m', so p is perpendicular to both l and m by Definition 3h. Since the polar of P' is unique, it follows that p' is the unique common perpendicular to l and m.

(b) Assume that l and m are two hyperbolic lines with a unique common perpendicular p. Then the corresponding projective lines l' and m' are both conjugate to p', the projective line corresponding to p, so they must intersect at the pole of p'. Denote this pole by P'. Since p' contains interior points of $\mathscr{C}$, p'

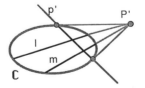

Figure 4.46

must intersect $\mathscr{C}$ at two distinct points, so P' is an ultraideal point. It follows that l and m are ultraparallel lines (Definition 2h). □

The definitions used for sensed-parallel and ultraparallel lines in hyperbolic geometry do not apply to single elliptic geometry (why?). However, the remaining three definitions do apply. Here, however, the points of intersection with $\mathscr{C}$ and the lines tangent to $\mathscr{C}$ referred to in Definitions 2e and 3e will necessarily have coordinates that involve complex numbers.

Single Elliptic Geometry

Ideal Points: Points of the absolute conic $\mathscr{C}$: $(x_1)^2 + (x_2)^2 + (x_3)^2 = 0$.
(Ordinary) Points: Points of the real projective plane.
Lines: Lines of the real projective plane.

Single Elliptic Definitions

1e: Two elliptic lines are *perpendicular* if the corresponding projective lines are conjugate with respect to $\mathscr{C}$.

2e: If A and B are two elliptic points, the *directed elliptic distance*, $d_e(A, B) = k \ln(R(A, B, P, Q))$ where P and Q are points of intersection of line AB with $\mathscr{C}$, and "ln" represents the natural logarithm.

3e: If a and b are elliptic lines than the *elliptic angle measure*, $m_e(\angle(a, b)) = k' \ln(R(a, b, p, q))$ where p and q are tangents to $\mathscr{C}$ from $a \cdot b$.

Again these definitions and theorems of projective geometry can be used to verify properties of the single elliptic plane. Since the absolute conic used to define the point set of the elliptic plane is empty, it is often not possible to depict properties of the single elliptic plane within the context of the projective plane. However, these properties can be illustrated within the model described in Section 2.9.

It is fairly easy to see that parallel lines can be defined in affine geometry using a definition analogous to that used for sensed-parallel lines in hyperbolic geometry. Since perpendicularity and angle measure are properties of similarity geometry, but not of affine geometry, it is appropriate to postpone their definitions until later. We can, however, include definitions of midpoint and types of conics in affine geometry.

Affine Geometry

Ideal Points: Points of the absolute conic $\mathscr{C}$: $x_3 = 0$.
(Ordinary) Points: Points of the real projective plane not on $\mathscr{C}$.
Lines: All lines of the real projective plane except the line $x_3 = 0$.

Affine Definitions

1a: Two affine lines are *parallel* if the corresponding projective lines intersect in ideal points.

2a: M is the *midpoint* of AB if $H(AB, PM)$ where P is the ideal point of line AB.

3a: A point conic is a *hyperbola, parabola,* or *ellipse* according to whether it contains two, one, or no real ideal points (see Fig. 4.47). The *center* of a conic is the pole of the ideal line with respect to the conic. The polar of any ideal point with respect to the conic is a *diameter* of the conic. A tangent to a conic at an ideal point is called an *asymptote*.

These definitions and theorems of projective geometry can be used to verify a number of affine properties, including the following.

Property 1a. *A line joining the midpoints of two sides of a triangle is parallel to the third side.*

Proof. Let M be the midpoint of AB in $\triangle ABC$ and let l be the unique parallel to BC through M (see Exercise 16). We shall show that $N = l \cdot AC$ is the midpoint of AC (see Fig. 4.48). Let $P = AB \cdot \mathscr{C}$. Then we have the harmonic set $H(AB, PM)$ by Definition 2a. Let $V = BC \cdot \mathscr{C}$; then $l = MV$. Finally, let $Q = AC \cdot \mathscr{C}$. We need to show that $H(AC, QN)$. This follows once we note that

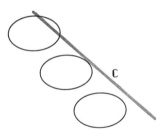

Figure 4.47

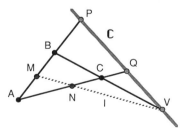

Figure 4.48

$ABPM \overset{V}{\barwedge} ACQN$, since harmonic sets are preserved under perspectivities. So N is the midpoint of AC.   □

In order to obtain similarity geometry, it is essential that we be able to give a definition for perpendicular lines. This was done in hyperbolic and single elliptic geometry using the absolute polarity. Since the absolute conic determining the affine and similarity planes is degenerate, there is no associated absolute polarity. However, we can introduce an absolute elliptic involution (i.e., a projectivity T such that $T^2 = I$) on $x_3 = 0$ to use in place of an absolute polarity. This absolute projectivity will have no invariant points since it is elliptic but will interchange pairs of points [i.e., if $T(P) = Q$, then $T(Q) = P$] since it is an involution. The projectivity we will choose will be the one that interchanges $X(1,0,0)$ and $Y(0,1,0)$.

In order to show that it is possible to use a definition of angle measure in similarity geometry like that used in the non-Euclidean geometries, it is necessary to determine what we mean by the line conic corresponding to our degenerate point conic $x_3 = 0$.

Recall that if a point conic has equation $X^t C X = 0$, then the corresponding line conic has equation $u C^{-1} u^t = 0$. In general, the absolute conic has matrix

$$C = \begin{bmatrix} c & 0 & 0 \\ 0 & c & 0 \\ 0 & 0 & 1 \end{bmatrix} \quad \text{with} \quad C^{-1} = \begin{bmatrix} 1 & 0 & 0 \\ 0 & 1 & 0 \\ 0 & 0 & c \end{bmatrix}$$

Hence the line conic associated with the point conic $c[(x_1)^2 + (x_2)^2] + (x_3)^2 = 0$ is $(u_1)^2 + (u_2)^2 + c(u_3)^2 = 0$. In the case under consideration, $c = 0$, so the corresponding degenerate line conic is $(u_1)^2 + (u_2)^2 = 0$. We can factor this as $(u_1 + iu_2)(u_1 - iu_2) = 0$ where $i^2 = -1$. The lines of this *line conic* are all lines through the *points* $I(i, 1, 0)$ and $J(i, -1, 0)$; any point P will be on two tangents to the absolute conic, namely, PI and PJ. Using these two lines, it is then possible to give a definition for angle measure comparable to that used in hyperbolic and single elliptic geometries. This definition, together with the definition of perpendicular lines, is stated later. The affine definitions given previously also apply in similarity geometry since it is defined in terms of the same absolute conic and has the same point set as affine geometry.

Similarity Geometry

Absolute Conic: $\mathscr{C}$: $x_3 = 0$.

Absolute Projectivity: The elliptic involution on $\mathscr{C}$ that interchanges $X(1,0,0)$ and $Y(0,1,0)$ and keeps $I(i,1,0)$ and $J(i,-1,0)$ invariant.

Ideal Points: Points of the absolute conic $\mathscr{C}$.

(Ordinary) Points: Points of the real projective plane not on $\mathscr{C}$.

Lines: All lines of the real projective plane except the line $x_3 = 0$.

Similarity Definitions

1s: Two lines of similarity geometry are *perpendicular* if their ideal points correspond under the absolute projectivity.

2s: If a and b are lines of similarity geometry, then the *angle measure*, $m_a(\angle(a, b)) = k\ln(R(a, b, p, q))$ where p and q are the tangents to $\mathscr{C}$ from $a \cdot b$, and "ln" represents the natural logarithm.

Using these definitions it is also possible to verify a familiar property of similarity geometry.

Property 1s. *Two lines* $u[u_1, u_2, u_3]$ *and* $v[v_1, v_2, v_3]$ *are perpendicular iff* $u_1v_1 + u_2v_2 = 0$.

Proof. The line $u[u_1, u_2, u_3]$ has the ideal point $U(-u_2, u_1, 0)$, which has homogeneous parameters $(-u_2, u_1)$ with respect to X and Y. Using a matrix representation relative to these base points, we can obtain U', the image of U under the absolute projectivity as follows (see Exercise 5):

$$sU' = \begin{bmatrix} 0 & 1 \\ -1 & 0 \end{bmatrix} U$$

So U' has homogeneous parameters (u_1, u_2) and therefore homogeneous coordinates $(u_1, u_2, 0)$. The lines perpendicular to u are those lines with U' as ideal point, but these lines are the lines $v[v_1, v_2, v_3]$ where $v_1u_1 + v_2u_2 = 0$. $\qquad\square$

It should be obvious after noticing the extra effort required to obtain the definitions for perpendicularity and angle measure for similarity and hence Euclidean geometry that it is more difficult to view Euclidean geometry as a subgeometry of projective geometry than it is to view the less familiar hyperbolic and single elliptic geometries this way. This added difficulty results since the absolute conic used to determine the point set for affine, similarity, and Euclidean geometry is a line rather than a conic determined by a polarity. In order to obtain similarity geometry, we had to introduce an elliptic involution on the ideal line. To obtain Euclidean geometry as a subgeometry of similarity geometry we could introduce a metric (i.e., a distance function). So Euclidean geometry can be described as a metric geometry based on an elliptic involution on the ideal line.

In closing, we will see that we can summarize the discussion in this section by applying one further adjective in the description of Euclidean geometry. This adjective describes a characterization based on the following comparison: (1) each line in the Euclidean plane has one real ideal point; (2) each line in the hyperbolic plane has two distinct real ideal points; and (3) each line in the single elliptic plane has no real ideal points. This should be reminiscent of the definition of types of point conics in the affine plane. There a conic is labeled a parabola, a hyperbola, or an ellipse depending on whether it contains one, two, or zero real ideal points. The two non-Euclidean geometries are thus appropriately called hyperbolic and elliptic. Likewise, Euclidean geometry can be classified as a *parabolic* geometry. The relations between these geometries are summarized in the following outline.

The Subgeometries of Plane Projective Geometry

Real Projective Geometry

Point set:

Points of real projective plane (equivalence classes from $\{x_1, x_2, x_3\}$)

Transformations:

Collineations

Hyperbolic Geometry	Affine Geometry	Single Elliptic Geometry
Points:	*Points*:	*Points*:
Points of real projective plane in interior of absolute conic $((x_1)^2 + (x_2)^2 - (x_3)^2 = 0)$	Points of real projective plane not on absolute conic $(x_3 = 0)$	All points of real projective plane
	Transformations:	
	Affinities	

Similarity Geometry

Points:

Points of affine plane

Transformations:

Similarities

Euclidean Geometry

Points:

Points of affine plane

Transformations:

Isometries

EXERCISES

1. Verify that the set of collineations that keep a given conic $\mathscr{C}$ invariant form a group.

2. Explain why ultraparallel lines cannot be defined in either affine or single elliptic geometry.

3. Verify that the collineations that keep $x_3 = 0$ invariant are the affinities described in Section 3.9.

4. If $A(0, 0, 1)$ and $B(1, 0, 1)$ are points of the affine plane, use Definition 2a to find the coordinates of the midpoint of AB.

5. Show that the matrix of the absolute projectivity used to define similarity geometry is

$$\begin{bmatrix} 0 & 1 \\ -1 & 0 \end{bmatrix}$$

[*Hint*: See Exercise 9 in Section 4.8.]

6. Show that the affinities which preserve the absolute projectivity used to define similarity geometry are the similarities described in Section 3.8. [*Hint*: If $S(P) = P'$ where S is the absolute projectivity with the matrix given in Exercise 5 and T is an affinity, find the conditions under which $S(T(P)) = T(P')$ for all points P on the ideal line.]

7. Show that similarities that are also equiareal transformations are the isometries of Euclidean geometry. (See Exercise 13 in Section 3.9.)

8. Show that in the case where $a \cdot b = Z(0, 0, 1)$ the definition of angle measure used in similarity geometry gives the same value as Definition 3.4 in Section 3.2 when $k = -i/2$ where $i^2 = -1$.

Use the definitions of the appropriate properties listed in this section, together with theorems from projective geometry, to verify each of the following results in hyperbolic, single elliptic, affine, and similarity geometries.

Hyperbolic Geometry

9. Any two points determine a unique line.

10. Any two distinct lines determine at most one point.

11. Through a given point P, not on a given line l, there are exactly two lines sensed parallel to l.

12. Through a given point there is exactly one line perpendicular to a given line.

Single Elliptic Geometry

13. Two points determine a unique line.

14. Two lines determine a unique point.

15. All the lines perpendicular to a given line are concurrent at a point, namely, the pole of the line.

Affine Geometry

16. Through a given point P not on a given line m there exists exactly one line parallel to m.

17. Two distinct lines parallel to the same line are parallel to each other.

18. If a line intersects one of two parallel lines, it intersects the other.

19. The medians of a triangle are concurrent. [*Hint*: Use perspective triangles.]

20. A hyperbola has a center that is an exterior point, an ellipse has a center that is an interior point, a parabola has a center on the absolute conic, and thus has no center in the affine plane.

21. Hyperbolas are the only conics with asymptotes in the affine plane. Parabolas have only a single asymptote, namely, the ideal line that is not in the affine plane.

22. The diameters of a conic go through the center.

23. A conic $X^t C X = 0$ is a hyperbola, parabola, or ellipse according to whether $(c_{12})^2 - c_{11}c_{22} >, =, < 0$.

Similarity Geometry

24. There is a unique line perpendicular to any given line through a given point.

25. A line perpendicular to one of two parallel lines is perpendicular to the other also.

26. Lines perpendicular to the same line are parallel to each other.

27. The altitudes of a triangle are concurrent. [*Hints*: First, recall that an altitude of $\triangle ABC$ is a line through vertex A and perpendicular to BC. Second, assume your triangle is $\triangle ABC$ where $A(0,0,1)$, $B(1,0,1)$, and $C(a,b,1)$.]

4.13. Suggestions for Further Reading

Coxeter, H.S.M. (1957). *Non-Euclidean Geometry*, 3rd ed. Toronto: University of Toronto Press. (Includes a detailed presentation of Euclidean and non-Euclidean geometries as subgeometries of projective geometry.)

Coxeter, H.S.M. (1961). *The Real Projective Plane*, 2nd ed. Cambridge: The University Press. (A primarily synthetic presentation restricted to the real plane, it includes the development of affine geometry.)

Coxeter, H.S.M. (1987). *Projective Geometry*, 2nd ed. New York: Springer-Verlag. (A classic text containing a detailed development of this geometry.)

Dorwart, H. (1966). *The Geometry of Incidence.* Englewood Cliffs, NJ: Prentice-Hall. (An expository overview of projective geometry.)

Kline, M. (1968). Projective geometry. In *Mathematics in the Modern World*: *Readings from Scientific American*, pp. 120–127. San Francisco: W.H. Freeman. (A short, easy-to-read introduction.)

Meserve, B.E. (1983). *Fundamental Concepts of Geometry.* New York: Dover. (Chapters 5 and 8 give a more detailed presentation of the material in Section 4.12.)

Mihalek, R.J. (1972). *Projective Geometry and Algebraic structures.* New York: Academic Press. (A detailed presentation emphasizing the interrelation between geometry and algebra.)

Pedoe, D. (1963). *An Introduction to Projective Geometry.* Oxford: Pergamon Press. (Contains an extensive treatment of the theorems of Desargues and Pappus.)

Penna, M.A., and Patterson, R.R. (1986). *Projective Geometry and Its Applications to Computer Graphics.* Englewood Cliffs, NJ: Prentice-Hall.

Seidenberg, A. (1962). *Lectures in Projective Geometry.* New York: Van Nostrand Reinhold. (The initial chapter introduces the major concepts in a fairly naive form; the remaining chapters develop the subject from axioms.)

Stevenson, F.W. (1972). *Projective Planes.* San Francisco: W.H. Freeman.

Tuller, A. (1967). *Modern Introduction to Geometries.* New York: Van Nostrand Reinhold. (Uses matrix representations of the projective transformations.)

Wylie, C.R. Jr. (1970). *Introduction to Projective Geometry.* New York: McGraw-Hill. (Contains both analytic and axiomatic developments.)

Young, J.W. (1930). *Projective Geometry.* The Carus Mathematical Monographs, No. 4. Chicago: Open Court Publishing Co. (for the M.A.A.). (Develops concepts intuitively first and then incorporates metric properties and group concepts.)

Readings on the History of Projective Geometry

Bronowski, J. (1974). The music of the spheres. In: *The Ascent of Man*, pp. 155–187. Boston: Little, Brown. This chapter is the companion to the 52-minute episode of the same name in *The Ascent of Man* television series.

Edgerton, S.Y. (1975). *The Renaissance Rediscovery of Linear Perspective.* New York: Basic Books.

Ivins, W.M. (1964). *Art and Geometry: A Study in Space Intuitions.* New York: Dover.

Kline, M. (1963). *Mathematics: A Cultural Approach.* Reading, MA: Addison-Wesley.

Kline, M. (1968). Projective geometry. In: *Mathematics in the Modern World*: *Readings from Scientific American*, pp. 120–127. San Francisco: W.H. Freeman.

Kline, M. (1972). *Mathematical Thought from Ancient To Modern Times.* New York: Oxford University Press.

Pedoe, D. (1983). *Geometry and the Visual Arts.* New York: Dover.

Suggestions for Viewing

The Art of Renaissance Science (1991, 45 min). Part IV relates the discovery and implementation of perspective in drawing and painting. Available from Science TV, P.O. Box 2498, Times Square Station, New York NY 10108.

Central Perspectivities (1971, 13.5 min). Demonstrates perspectivities and projectivities with flashing dots and lines. Produced by the College Geometry Project at the University of Minnesota. Available from International Film Bureau, 332 South Michigan Avenue, Chicago, IL 60604.

Masters of Illusion (1991, 30 min). Another illustration of the discovery and implementation of perspective in works of art. Produced and directed by Rick Harper, National Gallery of Art, Washington, DC.

Projective Generation of Conics (1971, 16 min). Illustrates four methods of constructing point conics and demonstrates their logical equivalence. Available from International Film Bureau, 332 South Michigan Avenue, Chicago, IL 60604.

Projective Geometry, Zeeman Masterclass Series with BBC (1986). Available from The Open University Production Centre, Walton Hall, Milton Keynes MK7 6BH, UK.

Euclid's Definitions, Postulates, and the First 30 Propositions of Book I*

Definitions

1. A *point* is that which has no part.
2. A *line* is breadthless length.
3. The *extremities of a line* are points.
4. A *straight line* is a line which lies evenly with the points on itself.
5. A *surface* is that which has length and breadth only.
6. The *extremities of a surface* are lines.
7. A *plane surface* is a surface which lies evenly with the straight lines on itself.
8. A *plane angle* is the inclination to one another of two lines in a plane which meet one another and do not lie in a straight line.
9. And when the lines containing the angle are straight, the angle is called *rectilineal.*
10. When a straight line set up on a straight line makes the adjacent angles equal to one another, each of the equal angles is *right,* and the straight line standing on the other is called *perpendicular* to that on which it stands.
11. An *obtuse angle* is an angle greater than a right angle.
12. An *acute angle* is an angle less than a right angle.
13. A *boundary* is that which is an extremity of anything.
14. A *figure* is that which is contained by any boundary or boundaries.
15. A *circle* is a plane figure contained by one line such that all the straight lines falling upon it from one point among those lying within the figure are equal to one another.
16. And the point is called the *centre* of the circle.
17. A *diameter* of the circle is any straight line drawn through the centre and terminated in both directions by the circumference of the circle, and such a straight line also bisects the circle.

*Reprinted with permission of Cambridge University Press from *The Thirteen Books of Euclid's Elements*, 2nd ed., pp. 154–155 (1956). Translated by Sir Thomas L. Heath. New York: Dover.

18. A *semicircle* is the figure contained by the diameter and the circumference cut off by it. And the centre of the semicircle is the same as that of the circle.
19. *Rectilineal* figures are those which are contained by straight lines, *trilateral* figures being those contained by three, *quadrilateral* those contained by four, and *multilateral* those contained by more than four straight lines.
20. Of trilateral figures, an *equilateral triangle* is that which has three sides equal, an *isosceles triangle* that which has two of its sides alone equal, and a *scalene triangle* that which has its three sides unequal.
21. Further, of trilateral figures, a *right-angled triangle* is that which has a right angle, an *obtuse-angled triangle* that which has an obtuse angle, and an *acute-angled triangle* that which has its three angles acute.
22. Of quadrilateral figures, a *square* is that which is both equilateral and right-angled; an *oblong* that which is right-angled but not equilateral; a *rhombus* that which is equilateral but not right-angled; and a *rhomboid* that which has its opposite sides and angles equal to one another but is neither equilateral nor right-angled. And let quadrilaterals other than these be called *trapezia*.
23. *Parallel* straight lines are straight lines which, being in the same plane and being produced indefinitely in both directions, do not meet one another in either direction.

The Postulates

1. To draw a straight line from any point to any point.
2. To produce a finite straight line continuously in a straight line.
3. To describe a circle with any centre and distance.
4. That all right angles are equal to one another.
5. That, if a straight line falling on two straight lines makes the interior angles on the same side less than two right angles, the two straight lines, if produced indefinitely, meet on that side on which are the angles less than the two right angles.

The Common Notions

1. Things which are equal to the same thing are also equal to one another.
2. If equals be added to equals, the wholes are equal.
3. If equals be subtracted from equals, the remainders are equal.
4. Things which coincide with one another are equal to one another.
5. The whole is greater than the part.

The First 30 Propositions of Book I

1. On a given finite straight line, to construct an equilateral triangle.
2. To place at a given point (as an extremity) a straight line equal to a given straight line.
3. Given two unequal straight lines, to cut off from the greater a straight line line equal to the less.
4. If two triangles have the two sides equal to two sides, respectively, and have the angles contained by the equal straight lines equal, they will also have the base equal to the base, the triangle will be equal to the triangle, and the remaining angles will be equal to the remaining angles, respectively, namely, those which the equal sides subtend.
5. In isosceles triangles, the angles at the base are equal to one another, and, if the equal straight lines be produced further, the angles under the base will be equal to one another.
6. If in a triangle two angles be equal to one another, the sides which subtend the equal angles will also be equal to one another.
7. Given two straight lines constructed on a straight line (from its extremities) and meeting in a point, there cannot be constructed on the same line (from its extremities), and on the same side of it, two other straight lines meeting in another point and equal to the former two, respectively, namely, each to that which has the same extremity with it.
8. If two triangles have the two sides equal to two sides, respectively, and have also the base equal to the base, they will also have the angles equal which are contained by the equal straight lines.
9. To bisect a given rectilinear angle.
10. To bisect a given finite straight line.
11. To draw a straight line at right angles to a given straight line from a given point on it.
12. To a given infinite straight line, from a given point which is not on it, to draw a perpendicular straight line.
13. If a straight line set up on a straight line make angles, it will make either two right angles or angles equal to two right angles.
14. If with any straight line, and at a point on it, two straight lines not lying on the same side make the adjacent angles equal to two right angles, the two straight lines will be in a straight line with one another.
15. If two straight lines cut one another, they make the vertical angles equal to one another.
16. In any triangle if one of the sides be produced, the exterior angle is greater than either of the interior and opposite angles.
17. In any triangle two angles taken together in any manner are less than two right angles.
18. In any triangle the greater side subtends the greater angle.
19. In any triangle the greater angle is subtended by the greater side.

20. In any triangle two sides taken together in any manner are greater than the remaining one.
21. If on one of the sides of a triangle, from its extremities, there be constructed two straight lines meeting within the triangle, the straight lines so constructed will be less than the remaining two sides of the triangle, but will contain a greater angle.
22. Out of three straight lines, which are equal to three given straight lines, to construct a triangle: thus it is necessary that two of the straight lines taken together in any manner should be greater than the remaining one.
23. On a given straight line and at a point on it, to construct a rectilineal angle equal to a given rectilineal angle.
24. If two triangles have the two sides equal to two sides, respectively, but have the one of the angles contained by the equal straight lines greater than the other, they will also have the base greater than the base.
25. If two triangles have the two sides equal to two sides, respectively, but have the base greater than the base, they will also have the one of the angles contained by the equal straight lines greater than the other.
26. If two triangles have the two angles equal to two angles, respectively, and one side equal to one side, namely, either the side adjoining the equal angles, or that subtending one of the equal angles, they will also have the remaining sides equal to the remaining sides and the remaining angle to the remaining angle.
27. If a straight line falling on two straight lines makes the alternate angles equal to one another, the straight lines will be parallel to one another.
28. If a straight line falling on two straight lines makes the exterior angle equal to the interior and opposite angle on the same side, or the interior angles on the same side equal to two right angles, the straight lines will be parallel to one another.
29. A straight line falling on parallel straight lines makes the alternate angles equal to one another, the exterior angle equal to the interior and opposite angle, and the interior angles on the same side equal to two right angles.
30. Straight lines parallel to the same straight line are also parallel to one another.

Hilbert's Axioms for Plane Geometry*

Undefined Terms: Point, line, plane, on, between, congruence.

Group I: Axioms of Connection

I-1. Through any two distinct points A, B, there is always a line m.

I-2. Through any two distinct points A, B, there is not more than one line m.

I-3. On every line there exist at least two distinct points. There exist at least three points which are not on the same line.

I-4. Through any three points, not on the same line, there is one and only one plane.

Group II: Axioms of Order

II-1. If point B is between points A and C, then A, B, C are distinct points on the same line, and B is between C and A.

II-2. For any two distinct points A and C, there is at least one point B on the line AC such that C is between A and B.

II-3. If A, B, C are three distinct points on the same line, then only one of the points is between the other two.

Definition. By the *segment AB* is meant the set of all points which are between A and B. Points A and B are called the *endpoints* of the segment. The segment AB is the same as segment BA.

II-4. **Pasch's Axiom.** Let A, B, C be three points not on the same line and let m be a line in the plane A, B, C which does not pass through any of the

*Reprinted with permission of Open Court Publishing Co. from D. Hilbert, *The Foundations of Geometry*, 2nd ed. (1921). Translated by E.J. Townsend. Chicago: Open Court Publishing Co.

points A, B, C. Then if m passes through a point of the segment AB, it will also pass through a point of segment AC or a point of segment BC.

Note: II-4′. This postulate may be replaced by the *separation axiom*. A line m separates the points of the plane which are not on m, into two sets such that if two points X and Y are in the same set, the segment XY does not intersect m, and if X and Y are in different sets, the segment XY does intersect m. In the first case X and Y are said to be on the *same side* of m; in the second case, X and Y are said to be on *opposite sides* of m.

Definition. By the *ray AB* is meant the set of points consisting of those which are between A and B, the point B itself, and all points C such that B is between A and C. The ray AB is said to *emanate from* point A.

A point A, on a given line m, divides m into two rays such that two points are on the same ray if and only if A is not between them.

Definition. If A, B, and C are three points not on the same line, then the system of three segments AB, BC, CA, and their endpoints is called the *triangle ABC*. The three segments are called the *sides* of the triangle, and the three points are called the *vertices*.

Group III: Axioms of Congruence

III-1. If A and B are distinct points on line m, and if A' is a point on line m' (not necessarily distinct from m), then there is one and only one point B' on each ray of m' emanating from A' such that the segment $A'B'$ is congruent to the segment AB.

III-2. If two segments are each congruent to a third, then they are congruent to each other. (From this it can be shown that congruence of segments is an equivalence relation; i.e., $AB \simeq AB$; if $AB \simeq A'B'$, then $A'B' \simeq AB$; and if $AB \simeq CD$ and $CD \simeq EF$, then $AB \simeq EF$.)

III-3. If point C is between A and B, and C' is between A' and B', and if the segment $AC \simeq A'C'$ and the segment $CB \simeq C'B'$, then segment $AB \simeq$ segment $A'B'$.

Definition. By an *angle* is meant a point (called the *vertex* of the angle) and two rays (called the *sides* of the angle) emanating from the point.

If the vertex of the angle is point A and if B and C are any two points other than A on the two sides of the angle, we speak of the angle BAC or CAB or simply of angle A.

III-4. If BAC is an angle whose sides do not lie on the same line and if in a given plane, $A'B'$ is a ray emanating from A', then there is one and only one ray $A'C'$ on a given side of line $A'B'$, such that $\angle B'A'C' \simeq \angle BAC$. In short, a given angle in a given plane can be laid off on a given side of a given ray in one and only one way. Every angle is congruent to itself.

Definition. If ABC is a triangle then the three angles BAC, CBA, and ACB are called the angles of the triangle. Angle BAC is said to be *included* by the sides AB and AC of the triangle.

III-5. If two sides and the included angle of the one triangle are congruent, respectively, to two sides and the included angle of another triangle, then each of the remaining angles of the first triangle is congruent to the corresponding angle of the second triangle.

Group IV: Axioms of Parallels (for a plane)

IV-1. **Playfair's Postulate.** Through a given point A not on a given line m there passes at most one line, which does not intersect m.

Group V: Axioms of Continuity

V-1. **Axiom of Measure (Archimedean Axiom).** If AB and CD are arbitrary segments, then there exists a number n such that if segment CD is laid off n times on the ray AB starting from A, then a point E is reached, where $n \cdot CD = AE$, and where B is between A and E.

V-2. **Axiom of Linear Completeness.** The system of points on a line with its order and congruence relations cannot be extended in such a way that the relations existing among its elements as well as the basic properties of linear order and congruence resulting from Axioms I–III and V-1 remain valid.

Note: V'. These axioms may be replaced by *Dedekind's axiom of continuity*. For every partition of the points on a line into two nonempty sets such that no point of either lies between two points of the other, there is a point of one set which lies between every other point of that set and every point of the other set.

Birkhoff's Postulates for Euclidean Plane Geometry*

Undefined Elements and Relations. (a) *Points*, $A, B, \ldots$; (b) sets of points called *lines*, $m, n, \ldots$; (c) *distance* between any two points: $d(A, B)$ a real nonnegative number with $d(A, B) = d(B, A)$; (d) *angle* formed by three ordered points $A, O, B, (A \neq O, B \neq O)$: $\angle AOB$, a real number (mod 2π). The point O is called the *vertex* of the angle.

Postulate I (Postulate of Line Measure). The points $A, B, \ldots$ of any line m can be put into 1:1 correspondence with the real numbers x so that $|x_B - x_A| = d(A, B)$ for all points A, B.

Definitions. A point B is *between* A and C $(A \neq C)$ if $d(A, B) + d(B, C) = d(A, C)$. The points A and C, together with all points B between A and C, form *segment AC*. The *half-line m'* with *endpoint O* is defined by two points O, A in line m $(A \neq O)$ as the set of all points A' of m such that O is not between A and A'. If A, B, C are three distinct points the three segments AB, BC, CA are said to form a *triangle* $\triangle ABC$ with *sides AB, BC, CA* and *vertices A, B, C*. If A, B, C are in the same line, $\triangle ABC$ is said to be *degenerate*.

Postulate II (Point-Line Postulate). One and only one line m contains two given points P, Q $(P \neq Q)$.

Definitions. If two distinct lines have no points in common they are *parallel*. A line is always regarded as parallel to itself.

Postulate III (Postulate of Angle Measure). The half-lines $m, n, \ldots$ through any point O can be put into 1:1 correspondence with the real numbers a (mod 2π) so that if $A \neq O$ and $B \neq O$ are points of m and n, respectively, the difference $a_n - a_m$ (mod 2π) is $\angle AOB$. Furthermore if the point B on n varies continuously in a line r not containing the vertex O, the number a_n varies continuously also.

Definitions. Two half-lines m, n through O are said to form a *straight angle* if $\angle mOn \equiv \pi$. Two half-lines m, n through O are said to form a *right angle* if $\angle mOn \equiv \pm \pi/2$, in which case we also say that n is *perpendicular* to m.

Postulate IV (Similarity Postulate). If in two triangles $\triangle ABC$, $\triangle A'B'C'$ and

---

*Reprinted with permission from G.D. Birkhoff, "A set of postulates for plane geometry (based on scale and protractor), *Annals of Mathematics* 33: 329–345 (1932).

for some constant $k > 0$, $d(A', B') = kd(A, B)$, $d(A', C') = kd(A, C)$ and $\angle B'A'C' = \pm \angle BAC$, then also $d(B', C') = kd(B, C)$, $\angle C'B'A' = \pm \angle CBA$, $\angle A'C'B' = \pm \angle ACB$.

Definitions. Any two geometric figures are *similar* if there exists a 1:1 correspondence between the points of the two figures such that all corresponding distances are in proportion and corresponding angles are all equal or all negatives of each other. Any two geometric figures are *congruent* if they are similar with $k = 1$.

APPENDIX D

The S.M.S.G. Postulates for Euclidean Geometry*

Undefined Terms: Point, line, plane.

Postulate 1. *Given any two different points, there is exactly one line which contains both of them.*

Postulate 2 (The Distance Postulate). *To every pair of different points there corresponds a unique positive number.*

Postulate 3 (The Ruler Postulate). *The points of a line can be placed in correspondence with the real numbers in such a way that:*

(i) *To every point of the line there corresponds exactly one real number.*

(ii) *To every real number there corresponds exactly one point of the line.*

(iii) *The distance between two points is the absolute value of the difference of the corresponding numbers.*

Postulate 4 (The Ruler Placement Postulate). *Given two points P and Q of a line, the coordinate system can be chosen in such a way that the coordinate of P is zero and the coordinate of Q is positive.*

Postulate 5. (a) *Every plane contains at least three noncollinear points.* (b) *Space contains at least four noncoplanar points.*

Postulate 6. *If two points lie in a plane, then the line containing these points lies in the same plane.*

Postulate 7. *Any three points lie in at least one plane, and any three noncollinear points lie in exactly one plane. More briefly, any three points are coplanar, and any three noncollinear points determine a plane.*

Postulate 8. *If two different planes intersect, then their intersection is a line.*

Postulate 9 (The Plane Separation Postulate). *Given a line and a plane containing it, the points of the plane that do not lie on the line form two sets such that:*

(i) *each of the sets is convex.*

(ii) *if P is in one set and Q is in the other then the segment $\overline{PQ}$ intersects the line.*

---

*Reprinted from S.M.S.G., *Geometry: Student's Text*, A.C. Vroman, Pasadena, CA, 1965.

Postulate 10 (The Space Separation Postulate). *The points of space that do not lie in a given plane form two sets such that:*

(i) *each of the sets is convex.*

(ii) *if P is in one set and Q is in the other, then the segment $\overline{PQ}$ intersects the plane.*

Postulate 11 (The Angle Measurement Postulate). *To every angle $\angle BAC$ there corresponds a real number between 0 and 180.*

Postulate 12 (The Angle Construction Postulate). *Let $\overrightarrow{AB}$ be a ray on the edge of the half-plane H. For every number r between 0 and 180 there is exactly one ray $\overrightarrow{AP}$, with P in H, such that $m\angle PAB = r$.*

Postulate 13 (The Angle Addition Postulate). *If D is a point in the interior of $\angle BAC$, then $m\angle BAC = m\angle BAD + m\angle DAC$.*

Postulate 14 (The Supplement Postulate). *If two angles form a linear pair, then they are supplementary.*

Postulate 15 (The S.A.S. Postulate). *Given a correspondence between two triangles (or between a triangle and itself), if two sides and the included angle of the first triangle are congruent to the corresponding parts of the second triangle, then the correspondence is a congruence.*

Postulate 16 (The Parallel Postulate). *Through a given external point there is at most one line parallel to a given line.*

Postulate 17. *To every polygonal region there corresponds a unique positive number.*

Postulate 18. *If two triangles are congruent, then the triangular regions have the same area.*

Postulate 19. *Suppose that the region R is the union of two regions R_1 and R_2. Suppose that R_1 and R_2 intersect at most in a finite number of segments and points. Then the area of R is the sum of the areas of R_1 and R_2.*

Postulate 20. *The area of a rectangle is the product of the length of its base and the length of its altitude.*

Postulate 21. *The volume of a rectangular parallelepiped is the product of the altitude and the area of the base.*

Postulate 22 (Cavalieri's Principle). *Given two solids and a plane, if for every plane which intersects the solids and is parallel to the given plane the two intersections have equal areas, then the two solids have the same volume.*

APPENDIX E

Some S.M.S.G. Definitions for Euclidean Geometry*

1. The *distance* between two points is the positive number given by the distance postulate (Postulate 2). If the points are P and Q, then the distance is denoted by PQ.
2. A correspondence of the sort described in Postulate 3 is called a *coordinate system* for the line. The number corresponding to a given point is called the *coordinate* of the point.
3. B is *between* A and C if (1) A, B, and C are distinct points on the same line, and (2) $AB + BC = AC$.
4. For any two points A and B the *segment* $\overline{AB}$ is the set whose points are A and B, together with all points that are between A and B. The points A and B are called the *endpoints* of $\overline{AB}$.
5. The distance AB is called the *length* of the segment $\overline{AB}$.
6. Let A and B be points of a line L. The *ray* $\overrightarrow{AB}$ is the set which is the union of (1) the segment $\overline{AB}$ and (2) the set of all points C for which it is true that B is between A and C. The point A is called the *endpoint* of $\overrightarrow{AB}$.
7. If A is between B and C, then $\overrightarrow{AB}$ and $\overrightarrow{AC}$ are called *opposite rays*.
8. A point B is called a *midpoint* of a segment $\overline{AC}$ if B is between A and C, and $AB = BC$.
9. The midpoint of a segment is said to *bisect* the segment. More generally, any figure whose intersection with a segment is the midpoint of the segment is said to *bisect* the segment.
10. The set of all points is called *space*.
11. A set of points is *collinear* if there is a line which contains all the points of the set.
12. A set of points is *coplanar* if there is a plane which contains all the points of the set.
13. A set A is called *convex* if for every two points P and Q of A, the entire segment $\overline{PQ}$ lies in A.
14. Given a line L and a plane E containing it, the two sets determined by Postulate 9 are called *half-planes*, and L is called an *edge* of each of them.

*Reprinted from S.M.S.G., *Geometry: Student's Text*, A.C. Vroman, Pasadena, CA, 1965.

We say that *L separates E* into the two half-planes. If two points *P* and *Q* of *E* lie in the same half-plane, we say that they lie *on the same side* of *L;* if *P* lies in one of the half-planes and *Q* in the other they lie on *opposite sides* of *L.*

15. The two sets determined by Postulate 10 are called *half-spaces,* and the given plane is called the *face* of each of them.

16. An *angle* is the union of two rays which have the same endpoint but do not lie in the same line. The two rays are called the *sides* of the angle, and their common endpoint is called the *vertex.*

17. If *A, B,* and *C* are any three noncollinear points, then the union of the segments $\overline{AB}, \overline{BC}, \overline{AC}$ is called a *triangle,* and is denoted by $\triangle ABC$; the points *A, B,* and *C* are called its *vertices,* and the segments $\overline{AB}, \overline{BC},$ and $\overline{AC}$ are called its *sides.* Every triangle determines three angles; $\triangle ABC$ determines the angles $\angle BAC, \angle ABC,$ and $\angle ACB,$ which are called the *angles* of $\triangle ABC.$

18. Let $\angle BAC$ be an angle lying in plane *E.* A point *P* of *E* lies in the *interior* of $\angle BAC$ if (1) *P* and *B* are on the same side of the line $\overleftrightarrow{AC}$ and (2) *P* and *C* are on the same side of the line $\overleftrightarrow{AB}$. The *exterior* of $\angle BAC$ is the set of all points of *E* that do not lie in the interior and do not lie on the angle itself.

19. A point lies in the *interior* of a triangle if its lies in the interior of each of the angles of the triangle. A point lies in the *exterior* of a triangle if it lies in the plane of the triangle but is not a point of the triangle or of its interior.

20. The number specified by Postulate 11 is called the *measure of the angle,* and is written as $m \angle BAC.$

21. If $\overrightarrow{AB}$ and $\overrightarrow{AC}$ are opposite rays, and $\overrightarrow{AD}$ is another ray, then $\angle BAD$ and $\angle DAC$ form a *linear pair.*

22. Angles are *congruent* if they have the same measure. Segments are *congruent* if they have the same length.

23. Given a correspondence $ABC \Leftrightarrow DEF$ between the vertices of the two triangles. If every pair of corresponding sides are congruent, and every pair of corresponding angles are congruent, then the correspondence $ABC \Leftrightarrow DEF$ is a *congruence between the two triangles.*

24. If the sum of the measure of two angles is 180, then the angles are called *supplementary,* and each is called a *supplement* of the other.

25. If the two angles of a linear pair have the same measure, then each of the angles is a *right angle.*

26. Two intersecting sets, each of which is either a line, a ray, or a segment, are *perpendicular* if the two lines which contain them determine a right angle.

27. The *perpendicular bisector* of a segment, in a plane, is the line in the plane which is perpendicular to the segment and contains the midpoint.

The A.S.A. Theorem

Theorem. *If in nondegenerate triangles* $\triangle ABC$ *and* $\triangle A'B'C'$, $\angle CAB \simeq \angle C'A'B'$, $\angle ABC \simeq \angle A'B'C'$, *and* $\overline{AB} \simeq \overline{A'B'}$, *then* $\angle BCA \simeq \angle B'C'A'$, $\overline{BC} \simeq \overline{B'C'}$, *and* $\overline{AC} \simeq \overline{A'C'}$.

The proof of this theorem is given in each of three axiom systems: Hilbert's Birkhoff's, and S.M.S.G.'s.

Proof I (based on Hilbert's axioms). We begin by proving the following.

Lemma. *If in nondegenerate triangles* $\triangle ABC$ *and* $\triangle A'B'C'$, $\overline{AB} \simeq \overline{A'B'}$, $\overline{AC} \simeq \overline{A'C'}$, *and* $\angle CAB \simeq \angle C'A'B'$; *then* $\overline{BC} \simeq \overline{B'C'}$, $\angle ABC \simeq \angle A'B'C'$, *and* $\angle BCA \simeq \angle B'C'A'$.

Proof. Let triangles $\triangle ABC$ and $\triangle A'B'C'$ be given as in the hypothesis of the lemma. Then $\angle ABC \simeq \angle A'B'C'$ and $\angle BCA \simeq \angle B'C'A'$ by Axiom III-5. Hence all that remains is to prove $\overline{BC} \simeq \overline{B'C'}$.

On ray $\overrightarrow{BC}$ find C'' such that $\overline{BC''} \simeq \overline{B'C'}$ (III-1), and construct AC'' (I-1) (Fig. F.1). If $C = C''$ then $\overline{BC''} \simeq \overline{BC}$ and hence $\overline{BC} \simeq \overline{B'C'}$.

If $C \neq C''$, the three points B, C, C'' are distinct since $\triangle ABC$ and $\triangle A'B'C'$ are nondegenerate triangles. Furthermore since C'' is on ray $\overrightarrow{BC}$ either C is between B and C'' or C'' is between B and C.

Case 1. C is between B and C''. Since $\overline{AB} \simeq \overline{A'B'}$, $\overline{BC''} \simeq \overline{B'C'}$, and $\angle ABC \simeq \angle A'B'C'$, $\angle C''AB \simeq \angle C'A'B'$ (III-5). But $\angle C'A'B' \simeq \angle CAB$.

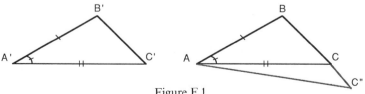

Figure F.1

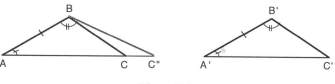

Figure F.2

Thus $\angle C''AB \simeq \angle CAB$.[†] So ray $\overrightarrow{AC}$ = ray $\overrightarrow{AC}$ " (III-4) and hence $\overline{AC''}$ = $\overline{AC}$ (I-1). Therefore C and C'' each lie on both AC and BC so $C = C''$. But this is a contradiction.

Case 2. C'' is between B and C. In this case a contradiction is obtained in a similar manner. Hence $C = C''$, and as a result $\overline{BC} = \overline{B'C'}$.

Let triangles $\triangle ABC$ and $\triangle A'B'C'$ be as given in the hypothesis. Then by the preceding lemma it suffices to show $\overline{AC} \simeq \overline{A'C'}$. On ray $\overrightarrow{AC}$ find C'' such that $\overline{AC''} \simeq \overline{A'C'}$ (III-1) and construct $\overline{BC''}$ (I-1). If $C = C''$, then $\overline{AC} = \overline{AC''}$ and hence $\overline{AC} \simeq \overline{A'C'}$. If $C \neq C''$, the remainder of the proof is analogous to the proof of the lemma.

Proof II (based on Birkhoff's axioms). Let triangles $\triangle ABC$ and $\triangle A'B'C'$ be as given in the hypothesis. Then by the definition of congruence $d(A, B) = d(A', B')$, $\angle CAB = \pm \angle C'A'B'$, $\angle ABC = \pm \angle A'B'C'$. We will assume that $\angle CAB = \angle C'A'B'$, $\angle ABC = \angle A'B'C'$ since the proof is similar in the other case. Using Postulate IV and the definition of congruence it suffices to show $d(A, C) = d(A', C')$. If $d(A, C) \neq d(A', C')$, then without loss of generality $d(A, C) < d(A'C')$. Let C'' be a point of ray $\overrightarrow{AC}$[‡] such that $d(A, C'') = d(A', C')$ (I), and consider $\overline{BC''}$ (II). Now $\angle C'A'B' = \angle CAB$. Furthermore since ray $\overrightarrow{AC}$ = ray $\overrightarrow{AC}$ " (II), $\angle CAB = \angle C''AB$ (III). Therefore $\angle C'A'B' = \angle C''AB$. In addition $d(A', C') = d(A, C'')$ and $d(A', B') = d(A, B)$. Hence $\angle A'B'C' = \angle ABC''$ (IV). This implies that $\angle ABC'' = \angle ABC$ so ray $\overrightarrow{BC}$ = ray $\overrightarrow{BC}$ " (III). Therefore C and C'' are each contained in BC and AC and so $C = C''$. It follows that $d(A, C) = d(A, C'')$. Since $d(A, C'') = d(A', C')$ this implies $d(A, C) = d(A', C')$.

Proof III (based on the S.M.S.G. axioms). Let triangles $\triangle ABC$ and $\triangle A'B'C'$ be as given in the hypothesis. By Postulate 15 it suffices to show $\overline{AC} \simeq \overline{A'C'}$. If $\overline{AC} \not\simeq \overline{A'C'}$, then by definition of congruent segments $AC \neq A'C'$. Without loss of generality, we can assume $AC < A'C'$. Let C'' be a point of line AC such that $\overline{AC''} \simeq \overline{A'C'}$ (Fig. F.2). Consider line BC'' (1). Now $m \angle CAB = m \angle C''AB$ (11) and $m \angle CAB = m \angle C'A'B'$ by the definition of congruent angles. Hence $m \angle C''AB = m \angle C'A'B'$; so $\angle C''AB \simeq \angle C'A'B'$. Furthermore $\overline{A'C'} \simeq \overline{AC''}$ by the definition of congruent segments, and $\overline{A'B'} \simeq \overline{AB}$ by hypothesis. Therefore $\angle ABC'' \simeq \angle A'B'C'$ (15); i.e. $m \angle ABC'' = m \angle A'B'C' = m \angle ABC$. This implies ray $\overrightarrow{BC}$ " = ray $\overrightarrow{BC}$ (12). And since line AC = line AC'' (1), it follows that C and C'' are each points of both lines AC and BC. Hence $C = C''$. Therefore $\overline{AC} \simeq \overline{AC''}$ so $\overline{AC} = \overline{AC''} \simeq \overline{A'C'}$.

[†]Note that we need to establish the transitivity of angle congruence.
[‡]Here it is necessary to show that this notation is well defined.

APPENDIX   G

References

Aaboe, A. (1964). *Episodes from the early History of Mathematics.* New York: L.W. Singer.

Adler, C.F. (1967). *Modern Geometry: An Integrated First Course.* New York: McGraw-Hill.

Adler, I. (1966). *A New Look at Geometry.* New York: John Day Co.

Adler, I. (1968). What shall we teach in high school geometry? *Mathematics Teacher* 61: 226–238.

Albert, A.A., and Sandler, R. (1968). *An Introduction to Finite Projective Planes.* New York: Holt, Rinehart and Winston.

Aleksandrov, A.D. (1969). Non-Euclidean geometry. In: A.D. Aledsandrov, A.N. Kolmogorov, and M.A. Lavrent'ev (Eds.), *Mathematics, Its Content, Methods, and Meaning,* Vol. 3, pp. 97–189. Cambridge, MA: M.I.T. Press.

Anderson, I. (1974). *A First Course in Combinatorial Mathematics.* Oxford: Clarendon Press.

Artzy, R. (1965). *Linear Geometry.* Reading, MA: Addison-Wesley.

Audsley, W.J. (1968). *Designs and Patterns from Historic Ornament.* New York: Dover.

Barker, S.F. (1964). *Philosophy of Mathematics.* Englewood Cliffs, NJ: Prentice-Hall.

Barker, S.F. (1984). Non-Euclidean geometry. In: D.M. Campbell and J.C. Higgins (Eds.), *Mathematics: People, Problems, Results,* Vol. 2, pp. 112–127. Belmont, CA: Wadsworth.

Beck, A., Bleicher, M.N. and Crowe, D.W. (1972). *Excursions into Mathematics.* New York: Worth Publishers.

Benedicty, M., and Sledge, F.R. (1987). *Discrete Mathematical Structures.* Orlando, FL: Harcourt Brace Jovanovich.

Birkhoff, G.D. (1932). A set of postulates for plane geometry, based on scale and protractor. *Annals of Mathematics,* 33: 329–345.

Blake, I.F., and Mullin, R.C. (1975). *The Mathematical Theory of Coding.* New York: Academic Press.

Blumenthal, L. (1961). *A Modern View of Geometry.* San Francisco: W.H. Freeman.

Bold, B. (1982). *Famous Problems of Geometry and How to Solve Them.* New York: Dover.

Borsuk, K. (1960). *Foundations of Geometry: Euclidean and Bolyai-Lobachevskian Geometry.* Amsterdam: North-Holland.

Bourbaki, N. (1950). The architecture of mathematics, *American Mathematical Monthly* 57: 221–232.

Boyer, C.B. (1956). *History of Analytic Geometry.* New York: Scripta Mathematica.

Bronowski, J. (1974). The music of the spheres. In: *The Ascent of Man,* pp. 155–187. Boston: Little, Brown.

Bruck R.H., and Ryser. H.J. (1949) The non-existence of certain finite projective planes. *Canadian Journal of Mathematics*. 1: 88–93.

Burn, R.P. (1975). *Deductive Transformation Geometry*. Cambridge: Cambridge University Press.

Cipra, B.A. (1988). Computer search solves an old math problem. *Science* 242: 1507–1508.

Copeland, R. (1979). *How Children Learn Mathematics: Teaching Implications of Piaget's Research*, 3rd ed. New York: Macmillan.

Courant, R., and Robbins, H. (1941). *What Is Mathematics?* London: Oxford University Press.

Coxeter, H.S.M. (1957). *Non-Euclidean Geometry*, 3rd ed. Toronto: University of Toronto Press.

Coxeter, H.S.M. (1961). *The Real Projective Plane*, 2nd ed. Cambridge: Cambridge University Press.

Coxeter, H.S.M. (1969). *Introduction to Geometry*, 2nd ed. New York: Wiley.

Coxeter, H.S.M. (1987). *Projective Geometry*, rev. 2nd ed. New York: Springer-Verlag.

Coxford, A.F., and Usiskin, Z.P. (1971). *Geometry: A Transformation Approach*. River Forest, IL: Laidlow Bros.

Dieudonné, J. (1981). The universal domination of geometry. *Two-Year College Mathematics Journal* 12: 227–231.

Dodge, C.W. (1972). *Euclidean Geometry and Transformations*. Reading, MA: Addison-Wesley.

Dorwart, H. (1966). *The Geometry of Incidence*. Englewood Cliffs, NJ: Prentice-Hall.

Dubnov, I.A. (1963). *Mistakes in Geometric Proofs*. Boston: Heath.

Eccles, F.M. (1971). *An Introduction to Transformational Geometry*. Menlo Park, CA: Addison-Wesely.

Edgerton, S.Y. (1975). *The Renaissance Rediscovery of Linear Perspective*. New York: Basic Books.

Eves, H. (1972). *A Survey of Geometry*, rev. ed. Boston: Allyn and Bacon.

Eves, H. (1976). *An Introduction to the History of Mathematics*, 4th ed. New York: Holt, Rinehart and Winston.

Ewald, G. (1971). *Geometry, An Introduction*. Belmont, CA: Wadsworth.

Faulkner, J.E. (1975). Paper folding as a technique in visualizing a certain class of transformations. *Mathematics Teacher* 68: 376–377.

Fehr, H.F., Eccles, F.M., and Meserve, B.E. (1972). The forum: What should become of the high school geometry course. *Mathematics Teacher* 65: 102ff.

Fey, J. (Ed.). (1984). *Computing and Mathematics: The Impact on Secondary School Mathematics*. Washington, DC: N.C.T.M.

Fishback, W.T. (1964). *Projective and Euclidean Geometry*. New York: Wiley.

Fisher, J.C. (1979). Geometry according to Euclid. *American Mathematical Monthly* 86: 260–270.

Gans, D. (1955). An introduction to elliptic geometry. *American Mathematical Monthly* 62 (7, part II): 66–73.

Gans, D. (1958). Models of projective and Euclidean space. *American Mathematical Monthly* 65: 749–756.

Gans, D. (1969). *Transformations and Geometries*. New York: Appleton-Century-Crofts.

Gans, D. (1973). *An Introduction to Non-Euclidean Geometry*. New York: Academic Press.

Gardner, M. (1959). Euler's spoilers: The discovery of an order-10 Graeco-Latin square. *Scientific American* 201 (5): 181–188.

Gardner, M. (1966). The persistance (and futility) of efforts to trisect the angle. *Scientific American* 214 (6): 116–122.

Gardner, M. (1975). On tessellating the plane with convex polygon tiles. *Scientific American* 233 (1): 112–117.

Gardner, M. (1978). The art of M.C. Escher. In: M. Gardner (Ed.), *Mathematical Carnival*, pp. 89–102. New York: Alfred A. Knopf.

Gardner, M. (1981). Euclid's parallel postulate and its modern offspring. *Scientific American* 254(4): 23–24.

Garner, L.E. (1981). *An Outline of Projective Geometry.* New York: Elsevier/North-Holland.

Gensler, H.J. (1984). *Gödel's Theorem Simplified.* Lanham, MD: University Press of America.

Golos, E. (1968). *Foundations of Euclidean and Non-Euclidean Geometry.* New York: Holt, Rinehart and Winston.

Gray, J. (1979). *Ideas of Space: Euclidean, Non-Euclidean, Relativistic.* Oxford: Clarendon Press.

Greenberg, M. (1980). *Euclidean and Non-Euclidean Geometries*, 2nd ed. San Francisco: W.H. Freeman.

Gruenberg, K.W., and Weir, A.J. (1967). *Linear Geometry.* New York: Van Nostrand Reinhold.

Grünbaum, B. (1981). Shouldn't we teach geometry? *Two-Year College Mathematics Journal* 12: 232–238.

Grünbaum, B., and Shephard, G.C. (1987). *Tilings and Patterns.* New York: W.H. Freeman.

Guggenheimer, H.W. (1967). *Plane Geometry and Its Groups.* San Francisco: Holden-Day.

Haak, S. (1976). Transformation geometry and the artwork of M.C. Escher. *Mathematics Teacher* 69: 647–652.

Hartshorne, R. (1967). *Foundations of Projective Geometry.* New York: W.A. Benjamin.

Heath, T.L. (1921). *A History of Greek Mathematics.* Oxford: Clarendon Press.

Heath, T.L. (1956). *The Thirteen Books of Euclid's Elements*, 2nd ed. New York: Dover.

Henderson, L.D. (1983). *The Fourth Dimension and Non-Euclidean Geometry in Modern Art.* Princeton, NJ: Princeton University Press.

Hilbert, D. (1921). *The Foundations of Geometry*, 2nd ed. E.J. Townsend (Trans.). Chicago: Opean Court Publishing Co.

Hilbert, D., and Cohn-Vossen, S. (1952). *Geometry and the Imagination.* P Nemenyi (Trans.). New York: Chelsea.

Hoffer, W. (1975). A magic ratio recurs throughout history. *Smithsonian* 6(9): 110–124.

Hofstadter, D.R. (1984). Analogies and metaphors to explain Gödel's theorem. In: D.M. Campbell and J.C. Higgins (Eds.), *Mathematics: People, Problems, Results*, Vol. 2, pp. 262–275. Belmont, CA: Wadsworth.

Iaglom, I.M. (1962). *Geometric Transformations*, Vols. 1, 2, 3. A. Shields (Trans.). New York: Random House.

Ivins, W.M. (1964). *Art and Geometry: A Study in Space Intuitions.* New York: Dover.

Jacobs, H. (1974). *Geometry.* San Francisco: W.H. Freeman.

Jeger, M. (1969). *Transformation Geometry.* English version by A.W. Deicke and A.G. Howson. London: Allen & Unwin.

Johnson, D.A. (1973). *Paper Folding for the Mathematics Class.* Washington, DC: N.C.T.M.

Kaplansky, I. (1969). *Linear Algebra and Geometry: A Second Course.* Boston: Allyn & Bacon.

Kelly, P., and Matthews, G. (1981). *The Non-Euclidean Plane: Its Structure and Consistency.* New York: Springer-Verlag.

Kennedy, H.C. (1972). The origins of modern axiomatics: Pash to Peano. *American Mathematical Monthly* 79: 133–136.

Klein, F. (1897). *Famous Problems of Elementary Geometry*. Boston: Ginn & Company.

Kline, M. (1963). *Mathematics: A Cultural Approach*. Reading, MA: Addison-Wesley.

Kline, M. (1964). Geometry. In: *Mathematics in the Modern World: Readings from Scientific American*, pp. 112–120. San Francisco: W.H. Freeman.

Kline, M. (1968). Projective geometry. In: *Mathematics in the Modern World: Readings from Scientific American*, pp. 120–127. San Francisco: W.H. Freeman.

Kline, M. (1972). *Mathematical Thought from Ancient to Modern Times*. New York: Oxford University Press.

Knorr, W.R. (1986). *The Ancient Tradition of Geometric Problems*. Boston: Birkhauser.

Kolata, G. (1982). Does Gödel's theorem matter to mathematics? *Science* 218: 779–780.

Lam, C.W.H. (1991). The Search for a Projective Plane of Order 10. *The American Mathematical Monthly*. Vol. 98, No. 4, pp. 305–318.

Lang, S., and Murrow, G. (1983). *Geometry: A High School Course*. New York: Springer-Verlag.

Lieber, L.R. (1940). *Non-Euclidean Geometry: Or Three Moons in Mathesis*, 2nd ed. Brooklyn, NY: Galois Institute of Mathematics and Art.

Lindquist, M.M., and Shulte, A.P. (1987). *Learning and Teaching Geometry, K–12: 1987 Yearbook*. Washington, DC: N.C.T.M.

Lockwood, E.H., and Macmillan, R.H. (1978). *Geometric Symmetry*. Cambridge: Cambridge University Press.

Lockwood, J.R., and Runion, G.E. (1978). *Deductive Systems: Finite and Non-Euclidean Geometries*. Reston, VA: N.C.T.M.

MacGillavry, C.H. (1976). *Symmetry Aspects of M.C. Escher's Periodic Drawings*, 2nd ed. Utrecht: Bohn, Scheltema & Holkema.

MacLane, S. (1959). Metric postulates for plane geometry. *American Mathematical Monthly* 66: 543–555.

Martin, G.E. (1982a). *The Foundations of Geometry and the Non-Euclidean Plane*, corrected ed. New York: Springer-Verlag.

Martin, G.E. (1982b). *Transformation Geometry: An Introduction to Symmetry*. New York: Springer-Verlag.

Maxwell, E.A. (1961). *Fallacies in Mathematics*. Cambridge: Cambridge University Press.

Maxwell, E.A. (1975). *Geometry by Transformations*. Cambridge: Cambridge University Press.

Maziarz, E., and Greenwood, T. (1984). Greek mathematical philosophy. In: D.M. Campbell and J.C. Higgins (Eds.), *Mathematics: People, Problems, Results*, Vol. 1, pp. 18–27. Belmont, CA: Wadsworth.

Meschkowski, H. (1964). *Non-Euclidean Geometry*, 2nd ed. A. Shenitzer (Trans.). New York: Academic Press.

Meserve, B.E. (1983). *Fundamental Concepts of Geometry*. New York: Dover.

Mihalek, R.J. (1972). *Projective Geometry and Algebraic Structures*. New York: Academic Press.

Mikami, Y. (1974). *The Development of Mathematics in China and Japan*, 2nd ed. New York: Chelsea.

Moise, Edwin E. (1974). *Elementary Geometry From An Advanced Standpoint*, Second Edition. Reading, MA: Addison-Wesley.

Nagel, E., and Newman, J.R. (1956). Goedel's proof. In: J.R. Newman (Ed.), *The World of Mathematics*, Vol. 3, pp. 1668–1695. New York: Simon and Schuster.

O'Daffer, P.G., and Clemens, S.R. (1976). *Geometry: An Investigative Approach*. Menlo Park, CA: Addison-Wesley.

Ogle, K.N. (1962). The visual space sense. *Science* 135: 763–771.

Olson, A.T. (1975). *Mathematics Through Paper Folding*. Washington, DC: N.C.T.M.

Osserman, R. (1981). Structure vs. substance: The fall and rise of geometry. *Two-Year College Mathematics Journal* 12: 239–246.

Pedoe, D. (1963). *An Introduction to Projective Geometry.* Oxford: Pergamon Press.

Pedoe, D. (1970). *A Course of Geometry for Colleges and Universities.* Cambridge: Cambridge University Press.

Pedoe, D. (1979). *Circles, A Mathematical View.* New York: Dover.

Pedoe, D. (1983). *Geometry and the Visual Arts.* New York: Dover.

Penna, M.A., and Patterson, R.R. (1986). *Projective Geometry and Its Applications to Computer Graphics.* Englewood Cliffs, NJ: Prentice-Hall.

Penrose, R. (1978). The geometry of the universe. In: L.A. Steen (Ed.), *Mathematics Today: Twelve Informal Essays,* pp. 83–125. New York: Springer-Verlag.

Piaget, J., and Inhelder, B. (1967). *The Child's Conception of Space.* F.J. Langdon and J.L. Lunzer (Trans.). New York: W.W. Norton.

Pless, V. (1982). *Introduction to the Theory of Error-Correcting Codes.* New York: Wiley.

Polya, G. (1971). *How to Solve It,* 2nd ed. Princeton, NJ: Princeton University Press.

Ranucci, E.R. (1974). Master of tessellations: M.C. Escher, 1898–1972. *Mathematics Teacher* 67: 299–306.

Robertson, J. (1986). Geometric constructions using hinged mirrors. *Mathematics Teacher* 79: 380–386.

Ryan, P.J. (1986). *Euclidean and Non-Euclidean Geometry: An Analytic Approach.* Cambridge: Cambridge University Press.

Sanders, W.J. and Dennis, J.R. (1968). Congruence geometry for junior high school. *Mathematics Teacher* 61: 354–369.

Sawyer, W.W. (1971). *Prelude to Mathematics.* New York: Penguin Books.

Schattschneider, Doris (1990). *M.C. Escher, Visions of Symmetry.* New York: W.H. Freeman and Company.

School Mathematics Study Group. (1965). *Geometry: Student's Text,* rev. ed. Pasadena, CA: A.C. Vroman.

Seidenberg, A. (1962). *Lectures in Projective Geometry.* New York: Van Nostrand Reinhold.

Smart, J.R. (1978). *Modern Geometries,* 2nd ed. Belmont, CA: Wadsworth.

Smith, D.E. (1958). *History of Mathematics,* Vol. 1. New York: Dover.

Solow, D. (1982). *How To Read and Do Proofs.* New York: Wiley.

Sommerville, D. (1970). *Bibliography of Non-Euclidean Geometry,* 2nd ed. New York: Chelsea.

Steen, L.A. (1980). Unsolved problems in geometry. *Mathematics Teacher* 73: 366–369.

Stevenson, F.W. (1972). *Projective Planes.* San Francisco: W.H. Freeman.

Swetz, F. (1984). The evolution of mathematics in ancient China. In: D.M. Campbell and J.C. Higgins (Eds.), *Mathematics: People, Problems, Results,* Vol. 1, pp. 28–37. Belmont, CA: Wadsworth.

Teeters, J.C. (1974). How to draw tessellations of the Escher type. *Mathematics Teacher* 67: 307–310.

Thompson, T.M. (1983). *From Error-Correcting Codes Through Sphere Packings to Simple Groups.* The Carus Mathematical Monographs, No. 21. Ithaca, NY: M.A.A.

Trudeau, R.J. (1987). *The Non-Euclidean Revolution.* Boston: Birkhäuser.

Tuller, A. (1967). *Modern Introduction to Geometries.* New York: Van Nostrand Reinhold.

Veblen, O., and Bussey, W.H. (1906). Finite projective planes. *Transactions of the American Mathematical Society* 7: 241–259.

Wenninger, M. (1966). *Polyhedron Models for the Classroom.* Washington, DC: N.C.T.M.

Wenninger, M. (1979). *Spherical Models.* Cambridge: Cambridge University Press.

Weyl, H. (1952). *Symmetry.* Princeton, NJ: Princeton University Press.

Whitehead, A.N. (1971). *The Axioms of Projective Geometry*. Cambridge Tracts in Mathematics and Mathematical Physics, No. 4. New York: Hafner Publishing Co.

Wolfe, H.E. (1945). *Introduction to Non-Euclidean Geometry*. New York: Holt, Rinehart and Winston.

Wylie, C.R., Jr. (1964). *Foundations of Geometry*. New York: McGraw-Hill.

Wylie, C.R., Jr. (1970). *Introduction to Projective Geometry*. New York: McGraw-Hill.

Yale, P.B. (1968). *Geometry and Symmetry*. San Francisco: Holden-Day.

Young, J.W. (1930). *Projective Geometry*. The Carus Mathematical Monographs, No. 4. Chicago: Open Court Publishing Co. (for the M.A.A.).

Zage, W.M. (1980). The geometry of binocular visual space. *Mathematics Magazine* 53:289–294.

Zirakzadeh, A. (1969). A model for the finite projective spaces with three points on every line. *American Mathematical Monthly* 76:774–778.

Index

Undergraduate Texts in Mathematics

(continued from page ii)

Malitz: Introduction to Mathematical Logic.

Marsden/Weinstein: Calculus I, II, III. Second edition.

Martin: The Foundations of Geometry and the Non-Euclidean Plane.

Martin: Transformation Geometry: An Introduction to Symmetry.

Millman/Parker: Geometry: A Metric Approach with Models. Second edition.

Moschovakis: Notes on Set Theory.

Owen: A First Course in the Mathematical Foundations of Thermodynamics.

Palka: An Introduction to Complex Function Theory.

Pedrick: A First Course in Analysis.

Peressini/Sullivan/Uhl: The Mathematics of Nonlinear Programming.

Prenowitz/Jantosciak: Join Geometries.

Priestley: Calculus: An Historical Approach.

Protter/Morrey: A First Course in Real Analysis. Second edition.

Protter/Morrey: Intermediate Calculus. Second edition.

Ross: Elementary Analysis: The Theory of Calculus.

Samuel: Projective Geometry. *Readings in Mathematics.*

Scharlau/Opolka: From Fermat to Minkowski.

Sigler: Algebra.

Silverman/Tate: Rational Points on Elliptic Curves.

Simmonds: A Brief on Tensor Analysis. Second edition.

Singer/Thorpe: Lecture Notes on Elementary Topology and Geometry.

Smith: Linear Algebra. Second edition.

Smith: Primer of Modern Analysis. Second edition.

Stanton/White: Constructive Combinatorics.

Stillwell: Elements of Algebra: Geometry, Numbers, Equations.

Stillwell: Mathematics and Its History.

Strayer: Linear Programming and Its Applications.

Thorpe: Elementary Topics in Differential Geometry.

Troutman: Variational Calculus and Optimal Control. Second edition.

Valenza: Linear Algebra: An Introduction to Abstract Mathematics.

Whyburn/Duda: Dynamic Topology.

Wilson: Much Ado About Calculus.

Princeton U-Store, NJ
Sunday 7 Dec 1997
$ 39.16 with 20% discount
+ 2.35 tax